Studies in Modern Chemistry

Courses in chemistry are changing rapidly in both structure and content. The changes have led to a demand for up-to-date books that present recent developments clearly and concisely.

This series is meant to provide chemistry students with books that will bridge the gap between the standard textbook and research paper. The books should also be useful to a chemist who requires a survey of current work outside his own field of research. Mathematical treatment has been kept as simple as is consistent with clear understanding of the subject.

Careful selection of authors actively engaged in research in each field, together with the guidance of four experienced editors, has ensured that each book ideally suits the needs of persons seeking a comprehensible and modern treatment of rapidly developing areas of chemistry.

Academic editor for this volume
T. C. Waddington, University of Durham

Studies in Modern Chemistry

R. L. M. Allen
Colour Chemistry

R. B. Cundall and A. Gilbert
Photochemistry

T. L. Gilchrist and C. W. Rees
Carbenes, Nitrenes, and Arynes

S. F. A. Kettle
Coordination Compounds

Ruth M. Lynden-Bell and Robin K. Harris
Nuclear Magnetic Resonance Spectroscopy

E. S. Swinbourne
Analysis of Kinetic Data

Martin L. Tobe
Inorganic Reaction Mechanisms

T. C. Waddington
Non-aqueous Solvents

K. Wade
Electron Deficient Compounds

Inorganic Reaction Mechanisms

Martin L. Tobe
University College London

Nelson

Thomas Nelson and Sons Ltd
36 Park Street London W1Y 4DE
P.O. Box 18123 Nairobi Kenya
Thomas Nelson (Australia) Ltd
597 Little Collins Street Melbourne 3000
Thomas Nelson and Sons (Canada) Ltd
81 Curlew Drive Don Mills Ontario
Thomas Nelson (Nigeria) Ltd
P.O. Box 336 Apapa Lagos
Thomas Nelson and Sons (South Africa) (Proprietary) Ltd
51 Commissioner Street Johannesburg

First published in Great Britain 1972
Copyright © Martin L. Tobe 1972
All Rights Reserved. No part of this publication may be reproduced, stored in a retrieval system, or transmitted, in any form or by any means, electronic, mechanical, photocopying, recording or otherwise, without the prior permission of the publishers.

Illustrations by Colin Rattray

ISBN 017 761720 9 (Boards)
     017 771719 X (Paper)

Printed offset in Great Britain by
The Camelot Press Ltd, London and Southampton

To the memory of Ron Nyholm

Contents

Preface *xi*

Introduction *1*

1 Determination of mechanism *4*
- 1-1 Introduction *4*
- 1-2 Structural information *4*
- 1-3 Kinetics *6*
- 1-4 Stability and inertness *6*
- 1-5 Rates and mechanism *7*
- 1-6 Dependence of rate on concentration *8*
- 1-7 Dependence of rate constant on the nature of the reagents *9*
- 1-8 Conclusion and moral *10*
  - Problems *10*
  - Bibliography *10*

2 Classification *11*
- 2-1 General remarks *11*
- 2-2 Reactions involving a change in the coordination shell *12*
- 2-3 Reactions involving a change in the oxidation state *12*
- 2-4 Reactions involving the ligands *13*
- 2-5 Apologia *15*
  - Problems *15*

3 Substitution reactions—general considerations *16*
- 3-1 Introduction *16*
- 3-2 Molecularity *18*
- 3-3 Classification *22*
- 3-4 Reactions leading to a change in coordination number *23*
  - Problems *24*
  - Bibliography *24*

4 Tetrahedral substitution *25*
- 4-1 Occurrence *25*
- 4-2 General features of substitution *27*
- 4-3 Light elements *27*
- 4-4 Stereochemistry of nucleophilic substitution at tetrahedral carbon *29*

viii Contents

    4-5 Heavier elements of the P-Block *32*
    4-6 Transition Elements *38*
         Problems *41*
         Bibliography *41*

5 Substitution at four-coordinate planar reaction centres *42*

    5-1 Occurrence *42*
    5-2 Coordination number and geometry associated with the d^8 configuration *44*
    5-3 Quasi-theoretical arguments *44*
    5-4 Direct kinetic evidence *45*
    5-5 Geometry of the transition states and intermediates *48*
    5-6 Factors controlling the reactivity of four-coordinate planar complexes *50*
    5-7 Unimolecular mechanism *63*
    5-8 Stereochemistry of substitution in planar four-coordinate complexes *66*
         Problems *68*
         Bibliography *68*

6 Substitution in five-coordinate systems *69*

    6-1 Introduction *69*
    6-2 Occurrence of five-coordination *69*
    6-3 General remarks on mechanism *71*
    6-4 Substitution in five-coordinate Ni(II), Pd(II), and Pt(II) complexes *73*
    6-5 Conclusion *74*
         Problems *75*
         Bibliography *75*

7 Substitution at six-coordinate reaction centres *76*

    7-1 Introduction *76*
    7-2 Quasi-theoretical arguments *77*
    7-3 General remarks on the distribution of experimental evidence *80*
    7-4 Kinetics and mechanism of the reactions of solvated metal cations *80*
    7-5 General kinetic features of octahedral substitution *83*
    7-6 Systematic discussion of the mechanism of octahedral substitution found in relatively inert systems *85*
    7-7 Unified view of the dissociative, D, and dissociative interchange, I_d, mechanism *100*
    7-8 Stereochemistry of octahedral substitution *103*
         Problems *106*
         Bibliography *107*

8 Stereochemical change *108*

8-1 Introduction *108*
8-2 General considerations of pseudorotation and consequent topological changes *111*
   Problems *122*
   Bibliography *123*

9 Oxidation and reduction *124*

9-1 Introduction *124*
9-2 Electron transfer *125*
9-3 Reactions of the solvated electron *125*
9-4 Outer sphere redox reactions *127*
9-5 Franck-Condon restrictions *129*
9-6 Inner sphere redox reactions *133*
9-7 Effect of non-bridging ligands *140*
9-8 Multiple bridging *141*
9-9 Differentiation between outer sphere and inner sphere mechanisms *142*
9-10 Number of electrons transferred *143*
9-11 Complementary and non-complementary reactions *145*
9-12 Catalysis of non-complementary reactions *148*
   Problems *150*
   Bibliography *150*

10 Redox addition, elimination, and substitution *151*

10-1 Introduction *151*
10-2 Oxidative addition *152*
10-3 Reductive elimination *162*
10-4 Redox substitution *163*
   Problems *166*

11 Catalysis and conclusions *168*

11-1 Catalysis *168*
11-2 Fixation of atmospheric nitrogen *169*
11-3 Polymerization of alkanes and alkynes *171*
11-4 Hydrogenation of alkanes *177*
11-5 Hydroformylation reaction *182*
11-6 What of the future? *184*
   Problems *187*
   Bibliography *187*

Index *189*

Preface

Judging by the increasing occurrence in examination papers of questions relating to one aspect or another of inorganic reaction mechanisms, one must assume that the subject forms an important part of the syllabus of inorganic chemistry even in departments where a large and active inorganic mechanisms research group is not operating. While there is no shortage of text books on the subject, many of these have been produced with the postgraduate in mind and so in writing this book I have tried to concentrate on what I personally would consider to be a suitable course at final year undergraduate level. The emphasis is very much on the inorganic side (as against the physico-chemical aspects of reaction mechanisms) but I have tried to show that carbon (and hence organic mechanisms) fits perfectly into the overall scheme even though the massive tail would wag the dog mercilessly if reaction mechanisms were not divided at the organic–inorganic borderline.

I have tried, where possible, to concentrate on patterns of behaviour and to put forward generalized views and so have had to avoid the 'ifs' and 'buts' that normally plague an honest book on reaction mechanisms. I have, nevertheless, tried to deal with facts and to illustrate the discussion with fact but this book has no pretentions towards being a review, comprehensive or otherwise. For this reason, no references to the original source of these facts are given. On the other hand, reference to further reading material is provided at the end of each chapter should anyone wish to read further into the subject. I make no excuse for having dwelt longest on those sections of the subject that interest me the most.

I would like to acknowledge my gratitude to the late Sir Christopher Ingold, who set me on the path of reaction mechanisms, and to the late Professor Sir Ronald Nyholm whose philosophy of inorganic chemistry has been borrowed extensively to provide the background theme of this book.

MARTIN L. TOBE.

Introduction

A major aim of modern inorganic chemistry is to understand the properties and behaviour, both physical and chemical, of inorganic substances in terms of their bonding, structure, and the mechanisms of their reactions, that is, to relate observations generally made on matter in bulk to the deduced description of the compound on the molecular level. The concept of valency—how and why atoms join together in chemical combination—has occupied the thoughts of chemists since before the days of Berzelius. The concepts of structure and symmetry—the three-dimensional arrangement of the bonded atoms—were developed initially in their modern form in the field of organic chemistry where Kekulé visualized the 'two-dimensional' ring structure of the aromatic nucleus and van'tHoff and Le Bel conceived the 'three-dimensional' tetrahedral disposition of the four bonds about a saturated carbon atom. Inorganic chemistry lagged behind considerably in this respect until Werner applied the concept of three-dimensional structure to what, at that time, was considered to be a group of unusual inorganic species: the so-called 'complex compounds'. These compounds could not be explained readily in terms of the then current theories of valency and Werner was able to show how these should be modified. One of the reasons why he was successful in demonstrating his theory was that he was using compounds that were able to retain their molecular integrity in solution whilst being subject to the experiments necessary to characterize their structure. This behaviour, which is normal for organic compounds, was quite unusual for the compounds of the metallic elements.

Shortly after this, the Braggs showed how the X-ray diffraction properties of crystals of ionic compounds could be interpreted in terms of the three-dimensional array of these ions, thereby making a further contribution to the concept of structure in inorganic chemistry. This technique, however, could not be applied at first to compounds that formed molecular or complex ionic crystals and any determination of their structure required the application of a range of chemical and physicochemical techniques, such as isomer counting, optical resolution, conductivity, dipole moments, and so on. These were all techniques that assumed a retention of molecular integrity in solution and the gas phase. Later, spectroscopic and magnetic studies were included in the technical armoury and coordination chemistry developed as the study of the preparation and properties of those compounds whose structures could most readily be deduced by available techniques.

2   Introduction

Meanwhile, the methods of interpreting X-ray diffraction of crystals were developed for molecular crystals, first for planar aromatic molecules and then for three-dimensional molecules both organic and inorganic. Nowadays, with automatic diffractometers and large computers, more and more inorganic structures are being determined with greater or lesser degree of reliability. Indeed, it is not uncommon to find published a structure of a compound whose preparation and properties have yet to be reported. It is likely that the indirect, or 'sporting' methods will eventually be used only for the study of structures in solution, where diffraction techniques are of little value.

As the concepts of structure and bonding have extended over the full range of inorganic chemistry, the distinction between 'simple' and 'complex' compounds has largely vanished and coordination chemistry has changed from a study of Werner-type complexes to an approach to the whole of chemistry.

Inorganic chemistry has also lagged behind organic chemistry in the determination of reaction mechanisms. The foundations and concepts of the mechanisms of reactions in solution were laid down and developed between 1920 and 1945 and applied almost entirely to the reactions of tetrahedral and trigonal planar carbon atoms in organic compounds. It is not difficult to see why carbon should have been the first reaction centre to be studied in such detail: (i) organic compounds often undergo reaction at one centre while all other bonds remain intact or else suffer only temporary change; (ii) the products of these reactions are kinetically controlled and so an indication of mechanism can be gained by a comparison of reagents and products; (iii) a great deal of information about interesting reactions was already available from preparative inorganic chemistry; (iv) the techniques for synthesizing compounds of known structure made it possible to prepare series of reagents in which a particular factor could be varied almost at will so that its effects upon the reaction could be used to aid any deduction of mechanism; (v) undoubtedly, the most important property was the slowness of reactions at a carbon atom so that conventional techniques of kinetics, probably the most useful tool for the study of mechanism, were readily applicable.

While organic chemistry in the 1920s presented fertile ground for the development of mechanism, inorganic chemistry did not. There was no background of systematized reactions and planned synthetic paths and preparative inorganic chemistry was, and to some extent still is, a rather haphazard and perhaps intuitive discipline. The view was widely held that virtually all inorganic reactions were either very rapid or else unselective so that the products were determined by thermodynamic and solubility considerations, unlike those of organic reactions which normally were determined by kinetic and hence mechanistic factors. In other words, the product of an inorganic reaction was thought to give little or no indication of its mechanism. Therefore, in spite of the great interest that was developed in the field of organic reaction mechanisms and the somewhat *avant-garde* views on substitution mechanisms that had been proposed by Werner in 1912, there were few published investigations of inorganic reactions in solution. Most of those dealing with metal complexes were motivated initially by problems far removed from mechanism such as salt effects, optical rotatory dispersion, stereochemical change, and the application of new or unusual methods to the study of reaction rates.

The great bulk of the information about inorganic reaction mechanisms that is available at the moment has been published in the past twenty years. At first, interest was concentrated on those areas where products were related to mechanism, and undue emphasis was placed on the behaviour of cobalt(III) and platinum(II) complexes, but in recent years the chemistry of other transition elements whose reactions are slow has been extensively studied. Mechanistic investigations have followed closely behind. The tremendous increase in interest in organo-metallic chemistry is leading to a new area of mechanistic importance and the breakdown of the distinctions between organic and inorganic chemistry. Indeed, the separation of organic and inorganic reaction mechanisms becomes artificial since the differences prove to be of emphasis and detail rather than of principle.

The development of readily available techniques for the measurement of fast reaction rates has allowed kinetic studies to be extended to the rapidly reacting systems and few areas of inorganic chemistry remain that are not amenable to some form of mechanistic study.

1 Determination of mechanism

1-1 Introduction

The concept of mechanism is at least one dimension more complicated than that of structure since it requires a detailed description of the way in which the structure and bonding of the reagents change with time in one individual act of chemical change.

In general terms, a complete description of the mechanism of a reaction would require:

(i) the subdivision of the reaction into its individual steps and equilibria;
(ii) the characterization of intermediate species and an estimate of their lifetime;
(iii) a description of the so-called 'transition state' for each reaction step in terms of (a) composition, (b) geometry, (c) solvation, and (d) energetics;
(iv) a complete description of the processes leading to and from each transition state in terms of the energy levels (electronic, vibrational, rotational, and so on) of the ground state and the excited states.

Such a complete description is well beyond the reach of present techniques, both practical and theoretical, except in the simplest of systems. It is always possible to deduce something about each of these factors but the extent and detail will depend upon the nature of the reaction and the quality of the techniques used.

Although this book will be mainly concerned with the deductions themselves it is instructive to revise, briefly, how such mechanistic deductions are made.

1-2 Structural information

An essential, but inadequate, part of the knowledge consists of the detailed structures of the reagents, intermediates, and products and the relationship between them. However, the techniques that lead to the determination of accurate structures only work for gases and crystalline solids. Solid state reactions are controlled almost entirely by the slow diffusion of reagents and products; reactions in the gas phase can be related to individual molecular collisions for the necessary transfer of energy and acts of chemical change. Reactions that take place in solution resemble those in the solid state in the sense that individual species are in contact with solvent molecules and therefore do not remain isolated until they collide but the rate of chemical change is often, but by no means always, considerably slower than the rate of diffusion. In the early days of studying reactions in solution the molecular nature of the environment was ignored and the solvent was treated as a dielectric continuum. This allowed the processes to be treated as quasi gas-phase reactions in which energy transfer (through the medium of the solvent) was always faster than any act of chemical change. This dangerous approximation was used with considerable success in the studies of typical organic reactions because the molecular nature of the solvent and its interaction with the reagents was only relevant when the finer details of mechanism were being considered. Unfortunately, for many inorganic reactions the molecular nature of the solvent cannot be ignored even in the first approximation and a detailed consideration of solvation, that is, the direct molecular interactions between solute and solvent is a fundamental requirement of any study of mechanism.

This then is the first, and possibly the most embarrassing handicap to the discussion of the mechanisms of reactions in solution. We might be fortunate and be able to describe the structure of the reagent, such as *trans*-$[Co\ en_2Cl_2]^+$ (en = 1,2-diaminoethane) by assuming that the complex ion retains its form on going into solution, but this retention of molecular integrity on going into solution is not a common feature of inorganic compounds. There is no way of using the information gained from the crystalline solid to describe the nature of the interaction between solute and solvent and other approaches and models must be adopted. The lack of precision in any of the available models for the liquid state and for solvent–solute interaction makes any discussion of solution chemistry of interesting systems in terms of mathematical expressions highly approximate. Anyone who expects to see reaction mechanisms in solution described by a sequence of mathematical equations will be disappointed at present. This state of affairs is welcome in the sense that the average chemist is not at the mercy of the mathematical experts and their true and false prophets, as we find in the quantum and wave-mechanical descriptions of chemical bonding. Unfortunately it carries the consequential disadvantage that the description is pictorial and semantic, and an impressive display of words and arm waving which may convince a captive audience may be shown to lack substance on further reflection.

The structural relationships may be further illuminated by the use of isotopic or optically active labels to identify the fate of various parts of the molecule.

1-3 Kinetics

The study of mechanism should really be the examination of the changing species *while it is actually changing*. It is not possible to determine a sequence of instantaneous structures of a molecule in the act of reaction and to string them together in a ciné-film and so, in order to understand the changing nature of the process, time-dependent phenomena must be studied. The rate of reaction is such a time-dependent process and kinetics, which is the area of chemistry relevant to reaction rates, provides a major tool for the assignment of mechanism. One should not confuse kinetics, which is the physicochemical tool, with mechanism, which is the interpretive description, but many people still do. A detailed discussion of kinetics, in terms of techniques, treatments, and computations is not within the scope of this book and many excellent textbooks on this subject are readily available. In order to see the relationship between reaction rate and mechanism it is necessary to consider reactivity in a little detail.

1-4 Stability and inertness

The concept of reactivity is of utmost importance in chemistry and yet the term itself is meaningless without further qualification. In order to bring sense to the subject it is convenient to consider separately the so-called thermodynamic and kinetic aspects of the problem.

The thermodynamic aspects of chemical reactions are concerned entirely with the beginning and the end and determine the theoretically possible extent of reaction, namely, the position of equilibrium, which will be determined by the difference in the standard free energies of the reagents and the products. The speed at which equilibrium is approached is a kinetic problem and relates directly to the mechanism of the reaction. Therefore, if we want to talk about the reactivity of a chemical compound we have to specify whether we are talking in a thermodynamic or a kinetic context. It has therefore become usual to reserve the term 'stability' (opposite = 'instability') for thermodynamic, that is, equilibrium, contexts and use the term 'inertness' (opposite = 'lability') for the kinetic aspects of reactivity. It is also meaningless to talk about the reactivity of a compound without either specifying or implying the reaction concerned. A few simple chemical examples may illustrate these points:

(i) NCl_3 is well recognized as a highly reactive species. It is unstable with respect to its elements, since the equilibrium

$$2NCl_3 \rightleftharpoons N_2 + 3Cl_2$$

lies completely over to the right. Its stability, in this context, does not depend upon its environment. When pure, it is inert, in the sense that it may be kept, if one is lucky (perhaps foolhardy would be a better word), indefinitely. However, the mechanism of its decomposition is such that shock or traces of catalytic materials may set off an explosive reaction. One could also discuss its reactivity towards other substrates in terms of equilibrium and mechanism.

(ii) $SiCl_4$ is also looked upon as a reactive substance but now the statement requires further qualification. It is stable with respect to its elements, that is, the equilibrium

$$SiCl_4 \rightleftharpoons Si + 2Cl_2$$

lies well over to the left and no amount of shock or catalysis will induce it to change spontaneously to silicon and chlorine. However, in the presence of water it is rapidly and irreversibly converted to hydrated silica and hydrochloric acid, that is,

$$SiCl_4 + \text{excess } H_2O \longrightarrow SiO_2(H_2O)_n + 4H^+_{aq} + 4Cl^-_{aq}$$

In this situation it is both unstable and labile.

(iii) CCl_4 can be shaken with water indefinitely and yet not change to carbon dioxide and hydrochloric acid even though the position of equilibrium should favour $CO_2 + HCl$. It is therefore unstable but quite inert towards hydrolysis. The difference in the hydrolytic behaviour of CCl_4 and $SiCl_4$ can be explained in terms of the mechanism of the reaction.

1-5 Rates and mechanism

The relationship between the rate of a reaction and its mechanism can be discussed at many levels of sophistication. In most cases, reaction will require the coming together of reagents and/or the separation of the product in individual molecular acts, but more than this is needed if the reaction is to take place. An individual reacting molecule or group of interacting species must achieve a configuration that is unstable with respect to its ground state and that of the products. This configuration is termed the 'transition state' and the individual reacting species must therefore gain and specially distribute excess energy in order to form the transition state and change into products. The overall process is indicated in Fig. 1-1 which is a somewhat fallacious, but nevertheless useful, two-dimensional representation of a single act of reaction. In general the process would have many degrees of freedom and would need a multi-dimensional surface to represent it. The free energy is plotted against the reaction coordinate, which in this figure represents changes in nuclear coordinates and electron distribution as the individual act of reaction progresses.

The weak point in this representation is that the concept 'free energy' can only be applied to matter in sufficient bulk for the statistics of energy distribution to be obeyed. The comparison of reagents and products in

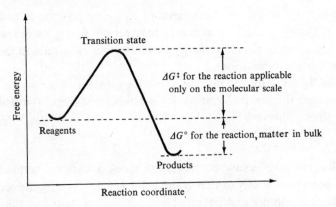

Fig. 1-1

terms of ΔG° and its relationship to the position of equilibrium is thermodynamically valid because we are relating to matter in bulk, but the actual journey through the transition state is done molecule by molecule. Transition state theory attempts to treat this as a macrochemical process. Thus it is not necessary to feed energy into the system (as is implied by the diagram) until $G°$ had increased to $G^{\ddagger}$ in order to achieve reaction, the distribution of the energy of individual species about the average is such that at any one time there may be a few reactant molecules that possess the right qualifications in position, energy, and energy distribution to allow them to react. Having qualified the facile diagrammatic representation in Fig. 1-1 in this way to avoid misunderstanding and to minimize (one can never fully eliminate) the criticism of the purists, we can then resume the simple approach and look upon the transition state as an energy barrier that the individual reacting species must cross in order to complete the reaction, so that, at a particular temperature, the rate of reaction will be determined by the height of the barrier. The dependence of rate upon the nature and the concentration of the reagents provides information about the composition of the transition state, although the interpretation of such data is not necessarily straightforward. The dependence of rate upon temperature allows the so-called free energy of activation to be divided up into an enthalpy and entropy of activation. It should be pointed out that precise identification of these quantities with true thermodynamic terms can be dangerous and often completely misleading. Nevertheless, we shall see that they are useful concepts if treated empirically.

1-6 Dependence of rate on concentration

The basic assumption is that the rate of reaction of a single stage is proportional to the concentrations of the species involved:

$$A \longrightarrow \text{products} \qquad \frac{-d[A]}{dt} = k_1[A]$$

or $A + B \longrightarrow \text{products} \qquad \dfrac{-dA}{dt} = \dfrac{-dB}{dt} = k_2[A][B]$

Unfortunately, nature is rarely kind enough to permit such a simple process and the actual act of chemical reaction is usually a sequence or network of such processes which we cannot readily examine in isolation. The observed rate law may then be complex, or more dangerously, deceptively simple. A major part of the text of a book on kinetics will be concerned with the unravelling of such expressions in terms of the rate constants for the individual processes.

An oft-quoted example is the reaction between hydrogen and bromine whose stoichiometric equation

$$H_2 + Br_2 \rightleftharpoons 2HBr$$

gives no indication of mechanism. The actual reaction consists of the sequence

$$Br_2 \underset{k_{-1}}{\overset{k_1}{\rightleftharpoons}} 2Br$$

$$Br + H_2 \underset{k_{-2}}{\overset{k_2}{\rightleftharpoons}} HBr + H$$

$$H + Br_2 \overset{k_3}{\longrightarrow} HBr + Br$$

which gives a kinetic expression

$$\frac{d[HBr]}{dt} = 2\left(\frac{k_1}{k_{-1}}\right)^{1/2} k_2[H_2][Br_2]^{1/2}\left(1 + \frac{k_{-2}[HBr]}{k_3[Br_2]}\right)$$

The agreement between the empirical expression and that derived above would indicate that the assigned sequence is not incorrect. This is not a complete description of the mechanism; it is just a start. It would still be necessary to describe, for example, the act whereby the bromine atom attacks a hydrogen molecule to form a hydrogen atom and a molecule of HBr.

1-7 Dependence of rate constant on the nature of the reagents

If we have been fortunate to derive a rate constant for a single stage of reaction then a great deal of information can be obtained from the way in which the rate constants, or the derived quantities such as $\Delta H^{\ddagger}$ or $\Delta S^{\ddagger}$, are affected by variation in the composition or structure of the reagents. This approach will be used many times in the main part of this book and

1-8 Conclusion and moral

In general, the assignment of mechanism is deductive. Once in a lifetime a particular experiment may produce compelling and unambiguous evidence for a particular mechanism but usually it is a question of finding a mechanism that is consistent with all the available facts, both kinetic and non-kinetic. It also has to fit in with current prejudice (or theory) otherwise no one will accept the paper. It also must be consistent with undiscovered facts (that is, it must be predictive). If, after finding 99 independent facts that are consistent with the mechanism, one is unfortunate to find one that is contradictory, the mechanism is wrong. The history of inorganic reaction mechanisms is filled with examples of mechanisms which, while perfectly valid within the available knowledge at the time, have had to be discarded or modified in the light of subsequent information. This method of progress is not restricted to the study of mechanism.

Problems

1-1 From your experience of preparative inorganic chemistry, write down four examples of reactions in which the products are determined kinetically and not thermodynamically.

1-2 The rates of exchange of free X^- with coordinated X in PtX_4^{2-} increase along the sequence $Cl < Br < I < CN$ and yet the thermodynamic stability of the PtX_4^{2-} complex increases in the same direction. Show that these two observations need not be contradictory.

1-3 What is the value of information gained from a study of the dependence of reaction rate upon temperature.

1-4 Write an essay on the limitations imposed upon the determination of mechanism when no kinetic data are available.

1-5 Write an account of the use of isotopes in the determination of mechanism.

Bibliography

Benson, S. W., *The Foundation of Chemical Kinetics*. McGraw-Hill, New York, 1960.
Frost, A. A. and R. G. Pearson, *Kinetics and Mechanism*, second edition. John Wiley, New York, 1961.
Harris, G. M., *Chemical Kinetics*. D. C. Heath, Chicago, 1966.

2 Classification

2–1 General remarks

It is proposed, in this book, to classify the different types of reactions in a way that is consistent with coordination chemistry. This approach to chemistry looks upon a chemical species in terms of a central atom, surrounded by a number of *ligands* that constitute the coordination shell. The coordination number (the number of ligands in the coordination shell), the geometry (the three-dimensional arrangement of the ligands about the central atom), the relative positions of the ligands within the coordination shell, and the oxidation state go a good way towards describing the immediate environment of the central atom. This approach is extremely useful in considering processes that can be focused at a single centre, and is therefore of considerable value in dealing with many reaction mechanisms. Since, in the first approximation, it makes no statement about the way in which the ligands are bound to the central atom, it can be applied validly to compounds ranging as widely as $NiCl_4^{2-}$, $FeCl_4^-$, $TiCl_4$, and CCl_4. It starts to lose its usefulness when we consider compounds or processes where there is no valid single focus, such as compounds with metal–metal bonds (whatever that term means); cluster compounds, for example, $Rh_4(CO)_{12}$; boranes, for example, $B_{10}H_{14}$; and most organic systems when there is no need to focus attention on one particular spot in the molecule.

For purposes of classification, inorganic reaction mechanisms can be divided into three major categories:

(1) reactions involving the coordination shell;
(2) reactions involving the oxidation state (Redox reactions);
(3) reactions involving the ligands.

But the situation is not always as simple as this and, even discounting complex multistage processes such as the reaction between MnO_4^- and $C_2O_4^{2-}$ which is made up of a large number of sequential reaction steps, there are many processes that involve a combination of two or more of these categories. It is, nevertheless, extremely useful to examine these categories separately at first and then consider the complications afterwards.

2-2 Reactions involving a change in the coordination shell

This category can be further subdivided as follows:

(a) increase of coordination number (addition);
(b) decrease of coordination number (elimination);
(c) ligand replacement (substitution);
(d) change in geometry;
(e) change of relative positions of ligands.

These subdivisions, while being extremely convenient, are arbitrary, and are closely interconnected. Thus, it is reasonably obvious that substitution must require a 'temporary' change in coordination number and we are left with the invidious task of defining the term 'temporary' if we wish to distinguish between (a) and (c) or (b) and (c). It is convenient to beg the question by restricting discussion of addition and elimination reactions to processes where the reagents and products are well defined chemical species. Thus, the gas phase reaction

$$(CH_3)_3B + N(CH_3)_3 \underset{\text{elimination}}{\overset{\text{addition}}{\rightleftharpoons}} (CH_3)_3BN(CH_3)_3$$

is a clear-cut example of addition (left to right) and elimination (right to left) and even in solution, the change in geometry from trigonal to tetrahedral boron allows us to accept that this is an addition and not a substitution.

A change in geometry, without breaking the metal–ligand bond, is quite a common phenomenon that is readily detected in a number of ways. For example, the planar $\rightleftharpoons$ tetrahedral interconversions of four-coordinate species, notably of Ni(II), where the planar species is diamagnetic and the tetrahedral species is paramagnetic, can be examined by the use of nuclear magnetic resonance.

A change in the relative positions of the ligands in the complex (stereochemical change) can be the result of substitution with steric change or it can be the consequence of a 'temporary' change of geometry. Thus process (e) can relate to (d) and (c) relates to (a) and (b).

2-3 Reactions involving a change in the oxidation state

Whereas, the concept of coordination number and coordination geometry is real, in the sense that it can be related to the distribution of electron density and nuclei that constitutes the structure, the concept of oxidation state is man-made and relates to an analysis of the complex in a way that does not necessarily represent a possible chemical reaction.

Thus the process

$$[Co(NH_3)_4NO_2Cl]^+ \longrightarrow Co^{3+} + 4NH_3 + NO_2^- + Cl^-$$

which is used to define the oxidation state of cobalt as +III, has no chemical reality whatsoever. Redox reactions can be subdivided as follows:

(a) electron transfer with no disturbance of coordination shells;
(b) processes requiring the sharing of a ligand (electron transfer through a bridge).

The least ambiguous redox process is one that takes place with complete retention of the coordination shell and simply involves a transfer of electrons from the reductant to the oxidant. Thus, in the reaction

$$[Fe(CN)_6]^{4-} + [Fe(phen)_3]^{3+} \longrightarrow [Fe(CN)_6]^{3-} + [Fe(phen)_3]^{2+}$$
$$Fe(II) \qquad Fe(III) \longrightarrow Fe(III) \qquad Fe(II)$$

it can be shown by suitable isotopic labelling that the coordination shells have remained intact throughout the process of electron transfer.

The formation of a bridge between oxidant and reductant can be demonstrated either by proving that a ligand is transferred as a result of the redox process or, in rarer cases, by isolating the actual bridged intermediate.

The problems arising from the artificial definition of oxidation state are stressed in those areas where the bonding is essentially covalent. Thus, is the process

$$O_3S^= + OCl^- \longrightarrow O_3SO^= + Cl^-$$
$$S(IV) + Cl(I) \longrightarrow S(VI) + Cl(-I)$$

a two-electron redox process, with simultaneous transfer of the bridging oxygen from chlorine to sulphur, or is it a nucleophilic attack by sulphite on oxygen, resulting in the displacement of chloride?

2-4 Reactions involving the ligands

This category comprises the most heterogeneous collection of phenomena but, in terms of its importance to the world at large (perhaps this is a useful definition of that overworked word 'relevance'), it is undoubtedly the most important. Often, but by no means always, the ligand is an organic species and the interest lies in the ways in which the organic reactions can be modified when the substrate is coordinated. All reactions which are either absent or grossly inefficient in the absence of a suitable centre for coordination are includable in this category.

Although a detailed discussion of this area of reaction type is beyond the scope of the book it is useful to indicate at this stage some of the types of reaction which individually or in combination allow processes which range from the use of organo-metallic complexes as specific catalysts for organic synthesis to the highly selective biochemical reactions of metallo-enzymes.

The following processes are probably the most important:

(a) *Site blocking.* This is a rather negative effect in which coordination serves to prevent reaction at the site favoured in the free ligand and allows an otherwise unobserved reaction to take place by default.

(b) *Polarization.* The metal ion can act as a temporary substituent that polarizes the reaction centre and favours reaction. This is a part played to perfection by the proton, but when subsequent attack by a basic reagent is necessary, a metal ion will allow simultaneous high concentrations of catalyst and reagent which the proton, as a result of acid–base effects, cannot allow.

(c) *Orientation.* By bringing reagents together in the coordination shell of the metal ion, modes of combination which are otherwise statistically unlikely can be promoted. This category would include the so-called 'template reaction' which can produce cyclic oligomers in a large variety of cases and the stereospecific polymerization of olefins.

(d) *Electron transfer.* The metal ion can act as a source or a sink of electrons and thereby reduce or oxidize a coordinated ligand. At other times the metal can act as a transmitter of electrons and allow the interaction of a pair of coordinated ligands. Oxidative addition, and its reverse reactions which will be discussed in a later chapter, falls into this category because it involves the change of a pair of electrons on the metal ion from a non-bonding to a bonding function. In this sense a ligand is reduced and can be split into two donor fragments. In general, the coordination number of the central atom increases on oxidation and both fragments can remain coordinated.

(e) *Ligand migration and insertion.* This is probably one of the most important functions of transition metal catalysts and involves processes whereby one ligand changes its mode of bonding, leaving a highly reactive centre to which another ligand in the coordination shell becomes attached. Such processes often require mediation of a third ligand to maintain the coordination number. Thus, the reaction

$$(CO)_5MnCH_3 + L \rightleftharpoons (CO)_4LMn(COCH_3)$$

involves the process

$$-\underset{CH_3}{\overset{|}{Mn}}-CO \rightleftharpoons -\underset{CH_3}{\overset{|}{Mn}}-\overset{O}{\underset{}{\overset{\|}{C}}} \underset{-L}{\overset{+L}{\rightleftharpoons}} -\underset{L}{\overset{|}{Mn}}-\overset{O}{\underset{CH_3}{\overset{\|}{C}}}$$

The change in the mode of bonding of the ligand is better seen in the reaction

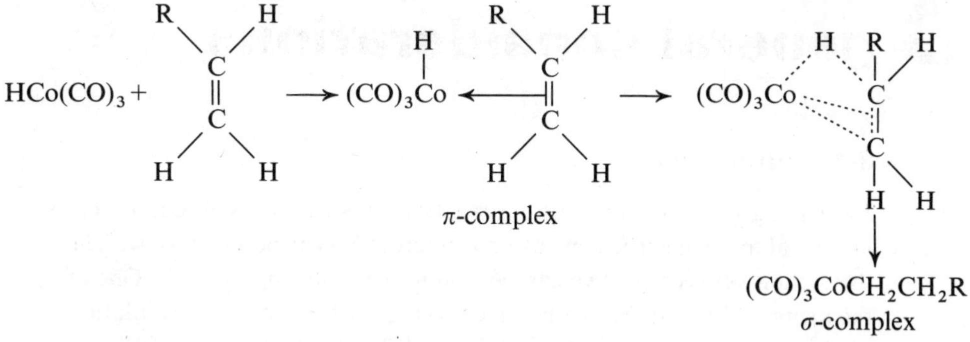

$$HCo(CO)_3 + \underset{H\quad H}{\overset{R\quad H}{\underset{C}{\overset{C}{\|}}}} \longrightarrow (CO)_3Co \longleftarrow \underset{H\quad H}{\overset{R\quad H}{\underset{C}{\overset{C}{\|}}}}-H \longrightarrow (CO)_3Co\cdots \underset{H\quad H}{\overset{H\ R\ H}{\underset{C}{\overset{C}{|}}}}$$

π-complex

$$\downarrow$$

$(CO)_3CoCH_2CH_2R$
σ-complex

which is a postulated step in the 'oxo' or hydroformylation reaction in which the combination of an olefin with CO and H_2 to form an aldehyde is catalysed by $Co_2(CO)_8$ or a similar catalyst.

2-5 Apologia

In a book of this size and scope it is quite impossible and impractical to cover everything. Rather than smear the butter thin over the whole loaf, the author proposes to concentrate on a limited number of areas which, as he would like to convince his reader (and himself), are the most important in the field. The cynic may be forgiven for taking the view that in this context the word 'important' is defined as meaning 'that which interests the author'. The topics covered will therefore be restricted to substitution reactions, intramolecular stereochemical change, redox reactions, and oxidative addition.

Problems

2-1 Classify the following reactions:
(a) cis-$[Co\ en_2(NH_2CH_2COOC_2H_5)Cl]^{2+} + OH^- \longrightarrow$
$[Co\ en_2(NH_2CH_2COO)]^{2+} + C_2H_5OH + Cl^-$
(b) $2FeCl_3 + SnCl_2 \longrightarrow 2FeCl_2 + SnCl_4$
(c) $[Co(CN)_5Cl]^{3-} + H_2O \longrightarrow [Co(CN)_5H_2O]^{2-} + Cl^-$
(d) trans-$[IrCl(CO)(PPh_3)_2] + HCl \longrightarrow [IrCl_2(H)(CO)(PPh_3)_2]$
(e) cis-$[Co\ en_2(OH)_2]^+ \longrightarrow$ trans-$[Co\ en_2(OH)_2]^+$
(f) $Os(bipy)_3^{2+} + Os(bipy)_3^{3+} \longrightarrow Os(bipy)_3^{3+} + Os(bipy)_3^{2+}$
(g) $BF_3 + (CH_3CH_2)_2O \longrightarrow F_3BO(CH_2CH_3)_2$
(h) $(C_6H_5)_3C^+BF_4^- + H_2O \longrightarrow (C_6H_5)_3COH + BF_4^- + H^+$
(i) $PtCl(CO)(PPh_3)_2^+ + CH_3O^- \longrightarrow PtCl(COOCH_3)(PPh_3)_2$
(j) $(+)-C_6H_{13}.\underset{\underset{CH_3}{|}}{C}HBr + Br^- \longrightarrow (-)-C_6H_{13}.\underset{\underset{CH_3}{|}}{C}HBr + Br^-$

3 Substitution reactions—general considerations

3-1 Introduction

In defining the term 'substitution reaction' it is necessary to choose between taking a macrochemical or a microchemical point of view. The distinction between the two can be seen in the following example. One of the recipes of Alfred Werner involves taking a saturated aqueous solution of the green *trans*-[Co en$_2$Cl$_2$]Cl (en = 1,2-diaminoethane, a bidentate ligand) and stirring it with a stick of sodium nitrite until it thickens to an orange paste of *cis*-[Co en$_2$NO$_2$Cl]Cl. On the macrochemical scale this is indeed a substitution reaction for we have succeeded in replacing a coordinated chloride by a nitro group. However, the reaction does not work if the reagents are pure and closer examination shows it to be an extremely complicated process involving, amongst other things, an oxidation-reduction path that requires the presence of Co(II).

In order to produce a definition that is mechanistically significant it is necessary to relate substitution to the molecular processes involved and to introduce the term '*simple substitution*'. This is defined as the replacement of a ligand in the coordination shell by another coming from the environment by a path that involves nothing more complicated than a temporary change in the coordination number of the reaction centre. The act of substitution is thus concerned with the making and breaking of bonds and a great deal of the description of mechanism relates to the changing electron distribution in the making and breaking bonds and also to the timing of the processes.

If we formulate the bonds in terms of shared pairs of electrons then it is possible to break the bond homolytically or heterolytically.

$$A:B \longrightarrow A\cdot + \cdot B \quad \text{Homolysis}$$

$$A:B \longrightarrow \overset{+}{A} + :B^- \quad \text{Heterolysis}$$

or $\quad A:B \longrightarrow \overset{-}{A}: + B^+ \quad \text{Heterolysis}$

A homolytic process in which each participant of the former bond takes one of the electrons will lead, when non-transition elements are involved, to unstable, odd-electron free radicals. With transition metal complexes, homolysis will lead effectively to the transfer of one electron, formerly involved in bonding, from the ligand to which it had been assigned by the rules of the game to the metal. The metal 'free radical' is likely to be no less stable than its precursor and the overall process is best looked upon as a reduction. Thus homolysis, which is a normal path of substitution in carbon chemistry, can become a redox process especially when applied to transition metal complexes.

A good example of this change of emphasis and nomenclature is the reaction between Cr_{aq}^{2+} and an organic halide, which is believed to proceed in two steps and generate an organic free radical as an intermediate.

$$C_6H_5CH_2Cl + Cr_{aq}^{2+} \longrightarrow (H_2O)_5CrCl^{2+} + C_6H_5CH_2^\bullet$$
$$\phantom{C_6H_5CH_2Cl + {}}Cr(II)Cr(III)\text{free radical}$$

$$Cr_{aq}^{2+} + C_6H_5CH_2^\bullet \longrightarrow (H_2O)_5CrCH_2C_6H_5^{2+}$$
$$Cr(II)Cr(III)$$

The reaction of the benzyl chloride is described in terms of a homolytic process whereas the complementary involvement of the Cr(II) species is looked upon as a one-electron oxidation. Homolytic reactions involving transition metal complexes are thus dealt with under the heading of redox reactions.

An exception to this is the bimolecular homolytic substitution (S_H2) process, which has recently been demonstrated elegantly by Davies, for example,

$$(CH_3)_3COOC(CH_3)_3 \xrightarrow{h\nu} 2(CH_3)_3CO^\bullet \quad \text{Homolysis}$$

$$(CH_3)_3CO^\bullet + B(CH_2CH_2CH_2CH_3)_3 \xrightarrow{S_H2}$$
$$(CH_3)_3COB(CH_2CH_2CH_2CH_3)_2 + {}^\bullet CH_2CH_2CH_2CH_3$$

This type of reaction has been found for a number of P-block reaction centres.

The heterolytic process may occur in one of two ways. The electrons of the bond may remain with the leaving group and the incoming group then needs to provide a replacing pair and seek a position of low electron density. Such a process is termed **nucleophilic** (S_N) and the **nucleophiles** are **Lewis bases**. Alternatively, the electrons of the bond may remain with the reaction centre and so the incoming group must act as an electron pair acceptor and seek a position of high electron density. Such reactions are termed **electrophilic**; **electrophiles** are **Lewis acids**. In the case of carbon, which has an intermediate electronegativity, both types of heterolysis are encountered and the process adopted will depend upon the nature of the ligands and the requirements of the reactions. Thus, an electronegative ligand such as Cl would be replaced nucleophilically from a tetrahedral carbon atom:

$$Nu{:}^- + R_3C-Cl \longrightarrow R_3C-Nu + {:}Cl^-$$

whereas an electropositive ligand, such as $HgCl^+$, would be replaced electrophilically:

$$R_3CHgCl + E^+ \longrightarrow R_3CE + HgCl^+$$

Complexes of transition metals, on the other hand, generally involve a reasonably electropositive centre, which are formulated as Lewis acids, to which are bonded a number of ligands formulated as Lewis bases. Ligand substitution is therefore a nucleophilic process. Electrophilic substitution reactions are rare, unless π-donation from the metal to the ligand plays an important part in the bonding of the ground state or the transition state. Reactions at 'metal–metal' bonds and even metal–carbon bonds may present new complications in this method of classification. Since bonds are made between atoms, the labels 'nucleophilic' and 'electrophilic' will depend upon the centre of attention. Thus the reaction represented as

$$(NH_3)_5CoCl^{2+} + H_2O \longrightarrow (NH_3)_5CoOH_2^{3+} + Cl^-$$

is clearly nucleophilic as we fully accept the cobalt atom as the reaction centre. However, when catalysed by Hg^{2+}

$$(NH_3)_5CoCl^{2+} + Hg^{2+} + H_2O \longrightarrow (NH_3)_5CoOH_2^{3+} + HgCl^+$$

it has sometimes been looked upon as an electrophilic process. Indeed, if we change the focus of our attention from cobalt to chlorine this is a correct description, but there is no real reason for us to do this.

3–2 Molecularity

The concept of molecularity of a reaction has caused, and still causes, considerable scope for discussion and polemic because it means different things to different people. A considerable amount of apparent disagreement arises from the varied forms of words used to describe what is essentially the same model of reaction.

In essence the molecularity of a reaction should give the necessary information about the composition of the transition state. In solution, a great deal of difficulty arises when one tries to take account of the change in solvation on going from the ground state to the transition state, and if this is included in the molecularity the concept becomes unwieldly almost to the point of uselessness. It was possibly for this reason that the following definition, most elegantly formulated by Ingold, was found to be acceptable. *The molecularity of a reaction stage is defined as the number of molecules necessarily undergoing covalency changes.* In this way solvation changes were relegated to a position of secondary importance. This is a valid approximation for most organic processes and for many inorganic reactions, but it does lead to problems when specific interaction with particular solvent molecules or even other solute species plays a significant role in the reaction even though the interacting species does not finish up bound to the reaction centre. We will see examples of this in the reactions of four-coordinate planar complexes where the reaction centre has a bare top and a bare bottom and also in the somewhat meaningless concept of the solvent-assisted dissociation of octahedral complexes. For nucleophilic substitution one might therefore talk of unimolecular (S_N1) and bimolecular (S_N2) substitutions (processes of higher molecularity being very rare indeed). The weakness of the above definition of molecularity is that it lacks any quantitative property and is not operational (that is, it is not linked unambiguously to a possible experimental evaluation). Occasionally there is a one to one correlation between molecularity and the kinetic order, that is, a bimolecular mechanism that has second-order kinetics, but the departures from this simplicity are so numerous that a somewhat more sophisticated approach is required.

An alternative view of molecularity, which applies solely to substitution reactions, relates to the fact that in such a process a bond is made and a bond is broken. The molecularity can be defined in terms of the timing of the bond-making and the bond-breaking aspects of substitution. Bond making and bond breaking can be synchronous or non-synchronous. A synchronous process takes place in one step and has one transition state and no intermediate. The transition state will determine both the energetics (rate) and stereochemistry of the reaction. The non-synchronous process can either have (i) bond breaking preceding bond making or (ii) bond making preceding bond breaking, and in both cases there will be an intermediate. If bond breaking comes first the intermediate will have a lower coordination number and the process is said to be **dissociative**. If bond making comes first the intermediate has a higher coordination number and the process is said to be **associative**. The simplified energy-reaction coordinate plots for these processes are shown in Fig. 3-1. A number of features emerge from this:

(i) The intermediate is not to be confused with the transition state.

(ii) When there is an intermediate there are two transition states, one for each stage of the substitution.

(iii) There is no restriction upon the lifetime of the intermediate. Generally it is extremely short—we are dealing with highly reactive intermediates—but this is not necessarily always the case and the description merges with that of a process leading to a change in coordination number.

The original S_N nomenclature has been modified to take account of the various processes. If there is definite evidence for an intermediate of lower coordination number the process is termed S_N1 lim. (lim. = limiting), and the extreme case of the associative mechanism when the

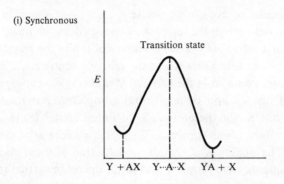

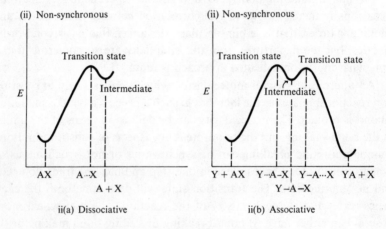

Fig. 3 – 1 Reaction profiles for substitution reactions

intermediate has virtually no bond weakening is termed S_N2 lim. and the rest is placed in between. This particular use of the nomenclature which appears to introduce a measure of quantification actually produces more harm than good because, while firmly describing the extremes, it leaves the intermediate region as a semantic battlefield.

Langford and Gray in their excellent monograph on substitution reactions attempted to produce an operational definition of mechanism. They introduce, first of all, the concept of stoichiometric mechanism and propose three simple pathways: (i) the dissociative (D) process, with an intermediate of lower coordination number; (ii) the associative (A) process, with an intermediate of higher coordination number; (iii) the interchange (I) process, in which the act of substitution is synchronous and no intermediate is involved. These categories are then modified by what is called the 'intimate mechanism' which relates to the activation of the process and can be either dissociative (**d**) or associative (**a**). An examination of Fig. 3-1 shows that for a dissociative process, the entering group does not interact directly with the reaction centre in the transition state, whereas for the associative process there is bonding. The experimental means of distinguishing the possible mechanisms then becomes possible. In a **d** (dissociatively activated) intimate mechanism the rate of reaction should be insensitive to the nature of the entering group (except for small effects arising from 'solvation' interactions), while the rate of an **a** (associatively activated) intimate mechanism should be very sensitive to the nature of the entering group. There are many kinetic and stereochemical techniques that can be used to detect the presence of reactive intermediates and these will be discussed in their context in later chapters. Such tests provide means for assigning an A or a D label which can only be done when evidence for the appropriate intermediate is forthcoming. The subscript labelling A_a and D_d is redundant since A mechanisms must be associatively activated and D mechanisms must be dissociatively activated. The interchange mechanism presents a greater problem, in the sense that its demonstration is negative, that is, an I mechanism is assigned when no evidence can be produced for a reactive intermediate. The interchange process can have either associative or dissociative activation and therefore I_a and I_d mechanisms are visualized. There is no difficulty in identifying the I_a mechanism with the synchronous bimolecular process (Fig. 3-1) but the I_d mechanism requires deeper thought. It will be shown in Chapter 8 that the D and the I_d process are best distinguished in terms of the lifetime of the intermediate. If it lives long enough to equilibrate its solvation

	Evidence for an intermediate of lower coordination number	No evidence for an intermediate		Evidence for an intermediate of higher coordination number
Stoichiometric mechanism	D	I		A
Intimate mechanism	D	I_d	I_a	A
Ingold definition	S_N1(Lim.)	S_N1	S_N2	S_N2(lim.)
Sensitivity of rate to nature of entering group	Rate independent of nature of entering group		Rate dependent on nature of entering group	

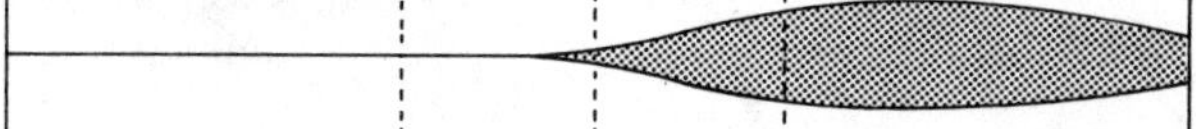

Fig. 3 – 2

environment it will obey the tests characteristic of a D mechanism. If it has to interact with the environment it has inherited, then the mechanism is I_d.

These various ways of describing the molecularity are very much interconnected and the relationship between them and the criteria of rate and so on are shown in Fig. 3-2.

3-3 Classification

If one thinks about the amount of work, the number of textbooks, the specializations and subdivisions of organic reaction mechanisms where the reaction centre, carbon, is just one of the 104 chemical elements (and a very well behaved one at that), the magnitude of the task facing us when we consider the other 90 odd feasible elements as reaction centres would appear to be overwhelming. Fortunately, at this time at least the information available, while being adequate, is by no means lavish. Doubly fortunate is the observation that, in spite of the scope for variation, the pattern of behaviour is extremely simple. Of the many possible systems of subdivision that are available—according to atomic number, position in the periodic table, electron configuration, type of bonding, oxidation state, coordination number, coordination geometry—the last two present the most useful basis from which to operate. It must be stressed that this is true for the situation at present and may very well change as the subject develops. This will be the basis for the subdivision adopted in this book and there appears to be a surprising degree of correlation between geometry and mechanism. Provided discussion is restricted to nucleophilic substitution reactions, the characteristic mechanistic features of the different geometries can be summarized in Table 3-1.

Table 3-1

Coordination number	Geometry	Characteristic mechanism
4	Tetrahedral	Depends on electron configuration e.g., D, I_d, I_a for C; I_a, A for Si, Ge; D, I_a, A for P; D, I_d for transition elements in low oxidation states. I_a, A for intermediate and high oxidation states.
4	Planar	A
5		Insufficient data for any systematic discussion.
6	Octahedral	D, I_d (I_a).
7 and higher		hardly studied.

3-4 Reactions leading to a change in coordination number

If one can imagine a modification of Fig. 3-1 in which the energies of the intermediate are lowered sufficiently to change them from highly reactive and unstable species to perfectly stable entities it would be possible to visualize a reaction in which there is a permanent change in coordination number. That in which there is an increase would be an **addition** reaction and that with a decrease would be an **elimination** reaction. A typical example is to be found in the reactions of suitable Pt complexes (Fig. 3-3). In this type of reaction, which is extremely rapid, it would be necessary to define the role of the axial solvent molecules S in order to distinguish between a substitution and an addition mechanism. A more clear cut example is to be found in boron chemistry (Fig. 3-4). Here the reaction can be studied in the gas phase and so the problem of solvation does not arise.

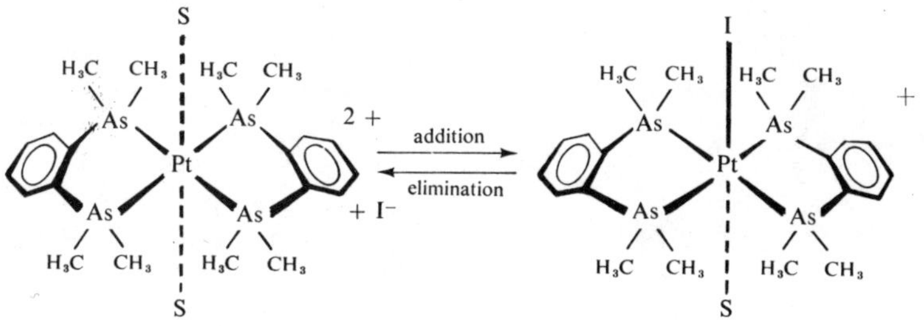

Fig. 3-3

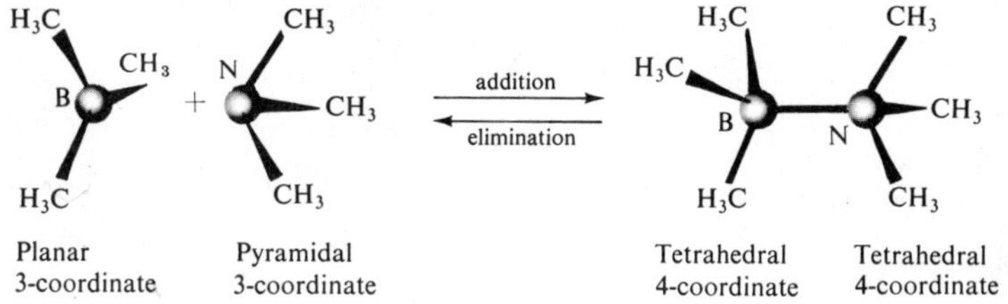

Planar
3-coordinate

Pyramidal
3-coordinate

Tetrahedral
4-coordinate

Tetrahedral
4-coordinate

Fig. 3-4

Problems

3-1 Define and distinguish the terms *order* and *molecularity* when applied to a substitution reaction.

3-2 Distinguish between homolytic and heterolytic modes of bond fission and discuss the circumstances under which each is favoured.

Bibliography

Basolo, F. and R. G. Pearson, *Mechanisms of Inorganic Reactions*, second edition. John Wiley, New York, 1967. This definitive work is relevant to all chapters in this book.

Langford, C. H. and H. B. Gray, *Ligand Substitution Processes*. W. A. Benjamin, New York, 1965.

4 Tetrahedral substitution

4-1 Occurrence

The tetrahedral geometry is extremely common and found in a very wide range of chemical species. The major areas are shown in Fig. 4-1. In terms of numbers of compounds, the field is completely dominated by the tetrahedral (aliphatic) carbon atom but this is just one prolific example of the characteristic behaviour expected for the light elements (Li → Ne) when there are four bonding pairs of electrons. The tetrahedral geometry is therefore common for boron and nitrogen in essentially covalent situations, although both can bear a lower coordination number with or without making use of multiple bonding. Lithium and beryllium likewise are characterized by a tetrahedral geometry but the bonding assumes an electrovalent character, especially in the case of lithium.

On going down the P-block to the heavier members of the groups there is scope for higher coordination numbers and indeed they tend to be favoured. Nevertheless, Al, Ga, Si, Ge, Sn, P, As, Cl, and I can all have tetrahedral compounds with the appropriate choice of ligand and some of these reaction centres have been studied in great detail.

In transition metal chemistry the tetrahedral geometry appears in a number of situations:

(i) Species of the type MX_4^{n-} ($n \geqslant 0$) in which X is usually a monodentate ligand such as chloride or bromide and the bonding is essentially electrovalent. The central ion, M, is either truly, or approximately spherical. Examples include $TiCl_4$, $MnCl_4^{2-}$, $FeCl_4^{-}$, $CoCl_4^{2-}$, $NiCl_4^{2-}$. This type of complex is rare for elements of the second and third rows of the D-block.

(ii) The oxy-anions and their simple derivatives, for example, MnO_4^{-}, CrO_4^{2-}, $Cr_2O_7^{2-}$.

(iii) Compounds such as $[Ni(PPh_3)_2X_2]$, where a combination of steric and electronic effects favours the lower coordination number and the high spin tetrahedral geometry.

(iv) Covalent compounds obeying the '18 electron rule', that is, d^{10} centres as in $Ni(CO)_4$, $Pt(PPh_3)_4$, and $Cu(Ph_3P)_3I$ and 'quasi d^{10} centres' such as $Co(NO)(CO)_3$.

The tetrahedral geometry is not common in lanthanide and actinide chemistry since these elements favour higher rather than lower coordination numbers.

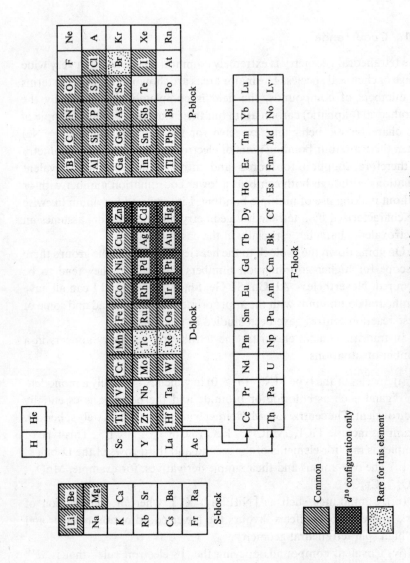

Fig. 4 – 1 Occurrence of tetrahedral four-coordination

4-2 General features of substitution

It is not possible in this context to talk about a typical tetrahedral substitution mechanism and each of the four groups listed above must be treated separately. It is possible to find whole textbooks dealing with homolytic and electrophilic substitution reactions at carbon in very great detail and many examples exist of other tetrahedral reaction centres undergoing these types of reactions, but in this book discussion will be confined almost entirely to nucleophilic processes.

4-3 Light elements

The pattern of reaction for this group is set by carbon and only this element will be discussed. Since it would be presumptuous to attempt to condense the many man-centuries of research into a few brief paragraphs it will only be possible to outline those aspects that are of interest in their application to the discussion of reactions at other centres. It is important to realize that there is no fundamental difference between inorganic and organic reaction mechanisms and any such subdivision is made purely for convenience.

The basic rules of the game played by carbon in its chemistry are extremely simple and an understanding of structure, bonding, and mechanism of the classical organic compounds—hydrocarbons and their derivatives with nitrogen, oxygen, and halogen donors—can be based on very simple valency theory which would get one nowhere if it had to be applied to, say, transition metal chemistry. We can see that this is true if we try to explain the analogous chemistry of the boranes and their derivatives or even the organo-metallic species such as ferrocene and other π-derivatives by the simple methods that work quite successfully for alkyl halides, ketones, carboxylic acids, and so on. The naivity and inherent conservatism of carbon with respect to its bonding habits is also reflected in its reaction patterns. No facile reaction path exists for substitution; indeed, all reaction paths are energetically most unfavourable, some more so than others. The result is that reactions at tetrahedral carbon are, under normal conditions, very much slower than the rate at which the reagents can come together. In addition, the rates of these reactions are extremely sensitive, not only to the nature of the leaving group, but also to the nature of the other ligands attached to the reaction centre and to the nature of the entering group when the reactions are associatively activated. This combination of kinetic inertness and sensitivity of reaction rate is probably the basis of the unique chemistry of carbon and the means whereby the vast range of compounds that constitute organic chemistry and, indeed, biology are able to exist.

Both homolytic and heterolytic substitution processes can be demonstrated for carbon. As a general rule, homolytic bond fission in an isolated molecule requires less energy than heterolytic fission since the latter leads to significant charge separation. Homolysis is therefore more likely in reactions that occur in the gas phase or in non-polar solvents. The extra energetics of charge separation in heterolysis can be adequately compensated by solvation in polar solvents and indeed heterolysis will be favoured in such solvents. There are many examples of heterolytic reactions in the gas phase and free radical (homolytic) reactions in polar solvents but these constitute the exception rather than the rule.

Heterolysis can be accomplished nucleophilically or electrophilically, depending upon the requirements of the reaction. As a rule, the standard substitution reactions of tetrahedral carbon atoms involve fairly electronegative entering and leaving groups and the processes are nucleophilic. Typical reactions are shown in Table 4-1.

There is no preponderant or characteristic mechanistic path for nucleophilic substitution. A dissociative (D) mechanism, with abundant evidence for a three-coordinate intermediate (carbonium ion) that reacts indepen-

Table 4-1 Typical substitution reactions of tetrahedral carbon

Nucleophilic

$(CH_3)_3CBr + H_2O \longrightarrow (CH_3)_3COH + H^+ + Br^-$	S_N1
$(CH_3)_2CHBr + H_2O \longrightarrow (CH_3)_2CHOH + H^+ + Br^-$	S_N1
$CH_3CH_2Cl + I^- \xrightarrow{acetone} CH_3CH_2I + Cl^-$	S_N2
$(CH_3)_3CCH_2Br + I^- \xrightarrow{acetone} (CH_3)_3CCH_2I + Br^-$	S_N2

Electrophilic

$sec\text{-}(C_4H_{10})_2Hg + HgBr_2 \longrightarrow 2\ sec\text{-}(C_4H_{10})HgBr$	S_E2
$CH_3HgBr + {}^*HgBr_2 \longrightarrow CH_3{}^*HgBr + HgBr_2$	S_Ei (S_E2 with a cyclic transition state)

$$\text{Ar-}CH_2Mn(CO)_5 + Hg^{2+} \longrightarrow \text{Ar-}CH_2Hg^+ + Mn(CO)_5^-$$

(where Ar = ortho-protonated pyridinium, $\overset{+}{N}H$)

S_E2 (open transition state)

$$HN^+\!\!-\!\!\bigcirc\!\!-\!\!CH_2Mn(CO)_5 + H_2O \longrightarrow HN^+\!\!-\!\!\bigcirc\!\!-\!\!CH_3 + Mn(CO)_5OH$$

S_E1 (S_N1 at manganese)

dently of its mode of formation, has been assigned to many reactions. In general, such a mechanism will be favoured by poor nucleophiles, labile leaving groups, and electron-releasing ligands attached to the reaction centre. In the limit, the carbonium ion can be generated under conditions where it cannot be consumed and it is then capable of being isolated, for example $(C_6H_5)_3C^+BF_4^-$. There has been extensive research on stable, unstable, and metastable carbonium ions, both the classical three-coordinate type considered here and the non-classical five-coordinate variety.

Bimolecular nucleophilic substitution is favoured by stronger nucleophiles and features that reduce the likelihood of a dissociative mechanism. Bond making and bond breaking contribute to the energetics of the transition state but since these usually have opposite requirements the response to other factors, such as electron displacement effects of the ligands and the strength of the bond with the leaving group, will depend entirely on the particular reaction. The mechanism is at all times I_a and no evidence for a reactive five-coordinate intermediate (let alone an isolatable five-coordinate species) has been forthcoming.

Electrophilic substitution is encountered when the leaving group is more electropositive than the carbon or when, under the influence of a strong electrophilic reagent, it can be persuaded to adopt, perhaps temporarily, an electron deficient role. The bimolecular mechanism has been well established for some time in certain *trans*-metallation reactions. At no stage is the carbon required to include more than eight electrons in its valence shell and an *A* mechanism may be considered to be possible. The bridging carbon in $Al_2(CH_3)_6$, if one believes the symmetrical structure, corresponds closely to a carbon atom undergoing associative electrophilic substitution. The dissociative process has only been demonstrated in recent times.

4-4 Stereochemistry of nucleophilic substitution at tetrahedral carbon

The relative simplicity of the mechanistic paths available to tetrahedral carbon is also reflected in the stereochemistry of substitution.

The bimolecular nucleophilic process is clearly defined. In order to accommodate the 10 electrons in the five-coordinate transition state a trigonal bipyramidal geometry is required with the entering and leaving groups in axial positions. In consequence, each act of bimolecular nucleophilic substitution at tetrahedral carbon leads to inversion of configuration (Fig. 4-2). The demonstration of this relationship in the late 1930s led to the full explanation of the Walden inversion. The dissociative process leads to the formation of a three-coordinate carbonium ion which, with six

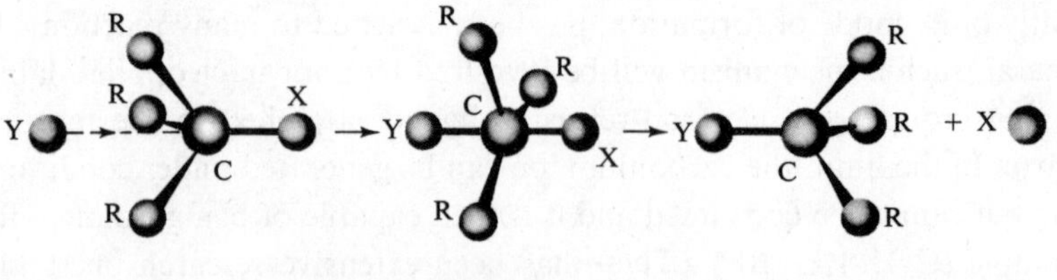

30 Tetrahedral substitution Ch. 4

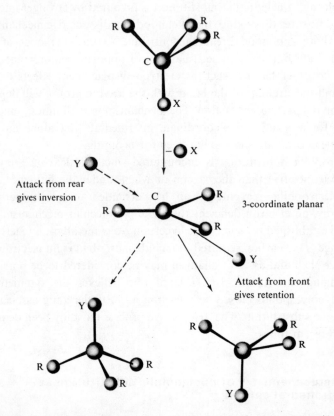

Trigonal bipyramidal transition state

Fig. 4 – 2 $S_N 2$ substitution at carbon leading to inversion of configuration

Attack from rear gives inversion

3-coordinate planar

Attack from front gives retention

Fig. 4 – 3 $S_N 1$ substitution at carbon with planar carbonium ion as intermediate

valence electrons, must adopt the planar configuration. The observed stereochemistry of unimolecular nucleophilic substitution should reflect this planar intermediate. As shown in Fig. 4-3, it would appear that the two paths along which the incoming group joins the reaction centre are fully equivalent and that equal quantities of the two potential enantiomeric products must form. However, the figure does not take account of the possibility that the leaving group is still sufficiently near to the carbonium ion to block approach adjacent to itself, nor does it take into account any

Table 4-2 Some steric courses of nucleophilic substitution at tetrahedral carbon

Substrate	Nucleophile	Product	Mechanism	% Retention	% Inversion
$(CH_3)(C_6H_{13})(H)CI$	*I$^-$	$(CH_3)(C_6H_{13})(H)C$*I	S_N2	0	100
$(CH_3)(C_6H_{13})(H)CCl$	$OC_2H_5^-$	$(CH_3)(C_6H_{13})(H)COC_2H_5$	S_N2	0	100
$(CH_3)(C_6H_{13})(H)CBr$	OH^-	$(CH_3)(C_6H_{13})(H)COH$	S_N2	0	100
$(CH_3)(C_6H_{13})(H)CBr$	H_2O	$(CH_3)(C_6H_{13})(H)COH$	S_N1	17	83
$(CH_3)(C_6H_{13})(H)CBr$	C_2H_5OH	$(CH_3)(C_6H_{13})(H)COC_2H_5$	S_N1	13	87
$(CH_3)(C_6H_5)(H)CCl$	OCH_3^-	$(CH_3)(C_6H_5)(H)COCH_3$	S_N2	0	100
$(CH_3)(C_6H_5)(H)CCl$	H_2O	$(CH_3)(C_6H_5)(H)COH$	S_N1	41	59
$(O_2C)(CH_3)(H)CBr^-$	OH^-	$(O_2C)(CH_3)(H)COH^-$	S_N2	0	100
$(O_2C)(CH_3)(H)CBr^-$	H_2O	$(O_2C)(CH_3)(H)COH^-$	S_N1	100	0

weak interaction between sites on the groups R and the cationic reaction centre. Table 4-2 gives a brief selection of steric courses for some nucleophilic reactions of tetrahedral carbon. The data in the table show clearly that the S_N2 process is invariably accompanied by complete inversion of configuration. The S_N1 reaction rarely leads to complete racemization in the act of substitution and the aliphatic secondary carbon atom in 2-octylhalides, for example, undergoes S_N1 substitution with predominant but not exclusive inversion of configuration. A phenyl substituent, on the other hand, causes the product of an S_N1 reaction to be closer to the 50:50 racemate, presumably because the lifetime of the three-coordinate intermediate is longer and the leaving group has virtually gone before the entering group is attached. A configurational holding substituent, such as COO^-, is electrostatically attracted to the developing carbonium ion before X has left and serves to block entry from the rear of the molecule. The product is then formed mainly with retention of configuration.

Signposting. The examination of the steric courses of substitution requires some signposting in order to relate macrochemical observation with the molecular process. The high symmetry of the tetrahedron makes this extremely difficult. If the reaction centre is restricted to an environment of four monodentate symmetrical ligands, then the only mode of signposting is to have all four different and make the reaction centre asymmetric (Fig. 4-4a). Much of the early work on the stereochemistry of tetrahedral substitution related to the relationship between mutarotation or racemization and the rates of substitution and it was necessary to know the relationship between the optical activity properties and the relative configuration. In cyclic systems, an inversion at one centre can be related to a signpost on an adjacent atom (Fig. 4-4b) and the stereochemistry can be examined by nuclear magnetic resonance techniques.

4-5 Heavier elements of the P-block

4-5-1 Silicon, germanium, and tin

The congeners of carbon in Group IVb, Si, Ge, Sn present many aspects of chemistry superficially similar to that of carbon. They differ mainly in the sense that coordination numbers of less than four are extremely rare while coordination numbers greater than four are quite common and, indeed, preferred as the group is descended. They are all considerably less electronegative than carbon, with the result that they invariably function as the Lewis acid component in any heterolytic process. These differences combine together to make a profound change in the mechanistic pattern of their substitution reactions. The most obvious change is the enormous increase in lability, so much so that to make adequate comparison fast-reaction techniques have to be employed for these compounds. Where the

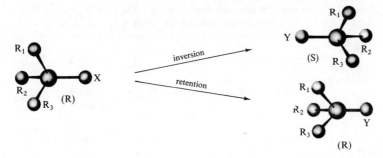

Fig. 4 – 4 (a) Signposting a monodentate tetrahedral system. It is necessary to provide an independent correlation of optical activity (sign of rotation, form of optical rotatory dispersion or circular dichroism) with the relative configurations of reagents and products

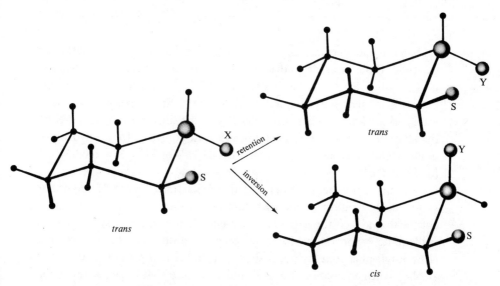

Fig. 4 – 4 (b) Signposting a chelated (or cyclic) tetrahedral system by a signpost (S) on an adjacent atom. This *cis* and *trans* forms can be distinguished and identified by their different n.m.r. spectra

kinetic order can be satisfactorily determined it is reasonably clear that the mechanism is bimolecular even under conditions most favourable to a dissociative mechanism. Thus $(C_6H_5)_3CCl$ will undergo substitution by a D mechanism and the triphenylmethyl carbonium ion can be isolated and characterized. $(C_6H_5)_3SiCl$, on the other hand, undergoes substitution in a typical bimolecular fashion. Where there is this mechanistic difference direct comparison of rate constants are meaningless. Thus, k_1 for the hydrolysis of $(C_6H_5)_3MF$ in 40% aqueous acetone at 25° is $2 \cdot 1 \times 10^{-4}$ s^{-1} for M = C and $1 \cdot 5 \times 10^{-6}$ s^{-1} for M = Si. However, while added base has no effect on the carbon system, which reacts dissociatively, it can increase the rate of reaction in the associative silicon system by a factor of 10^6.

Table 4-3 First-order rate constants for the reaction $R_3MCl + {}^{36}Cl^- \overset{k}{\rightleftharpoons} R_3M^{36}Cl + Cl^-$ in 2:5v/v acetone-dioxane at 25° (LiCl = 10^{-3}M)

$M = Si$						
R	C_2H_5	$n\text{-}C_3H_8$	$n\text{-}C_4H_{10}$	$n\text{-}C_6H_{14}$	C_6H_5	$C_6H_5CH_2$
$10^3 k(s^{-1})$	38	18	11	9	0·9	400
R	$C_6H_5CH_2CH_2$		cyclo-C_6H_{12}		1-naphthyl	
$10^3 k(s^{-1})$	140		0·007		0·0013	
$M = Ge$						
R	$n\text{-}C_4H_{10}$		$n\text{-}C_6H_{14}$	C_6H_5	cyclo-C_6H_{12}	
$10^3 k(s^{-1})$	1300		920	150	40	
$M = Sn$						
R	cyclo-C_6H_{12}					
$10^3 k(s^{-1})$	>4000					

Silicon and germanium present a similar mechanistic pattern and reactivity differences depend upon the nature of the reagents. The data in Table 4-3, for the chloride exchange of R_3MCl in acetone–dioxane, indicates a reactivity sequence, Si < Ge ≪ Sn. Nevertheless many examples are known where the reactivity of the silicon analogue is greater than that of germanium. For example, the rate constants for the hydrolysis of Ph_3MCl in acetone by $4M$ water are 4·03 s^{-1} for M = Si and 0·034 s^{-1} for M = Ge.

The distinction between an I_a and an A mechanism for these bimolecular substitutions would rest upon whether or not there was kinetic or stereochemical evidence for a reactive five-coordinate intermediate. There is inadequate information for any generalization but the report that (1-napthyl)-phenylmethylfluorosilane racemizes in n-pentane and t-butyl alcohol when methanol is added and at a rate that depends upon the concentration of methanol suggests that a symmetrical intermediate of higher coordination number is formed, although, in this case, cleavage of the Si—F bond does not follow (Fig. 4-5). A similar reaction has been observed for the analogous chlorogermane.

The stereochemical rules associated with substitution are much less rigid than in the case of tetrahedral carbon. Inversion of configuration is observed in a number of cases and retention of configuration in others. The rate of chloride exchange with Np.PhMeSiCl (Np = 1-naphthyl) in 5:1 dioxane-acetone is equal to the rate of racemization (not half the rate as would be required for substitution with inversion) and a symmetrical five-coordinate intermediate is indicated. The stereochemical behaviour seems to depend upon the requirements of the leaving group, X. A good leaving group such as Cl will generally be replaced with inversion of configuration and a trigonal bipyramidal intermediate or transition state with entering and leaving groups in axial positions is indicated. On the

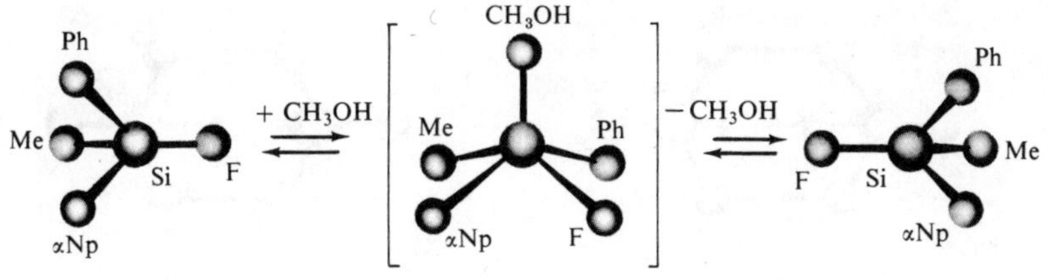

Fig. 4–5 Suggested path for the racemization of NpPhMeSiF in the presence of methanol

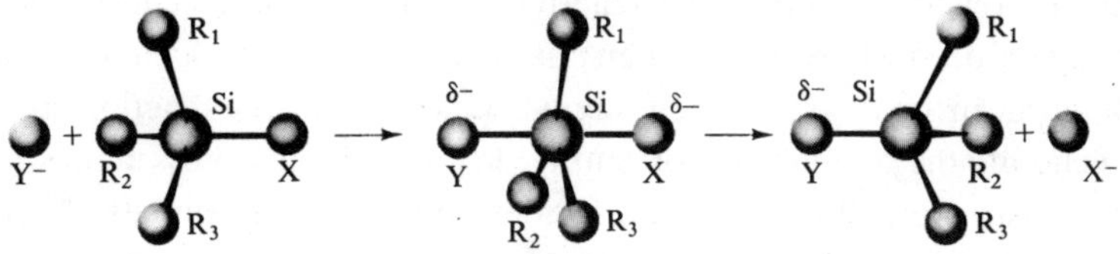

(a) X is a good leaving group—substitution with inversion by way of a trigonal bipyramid with X and Y axial

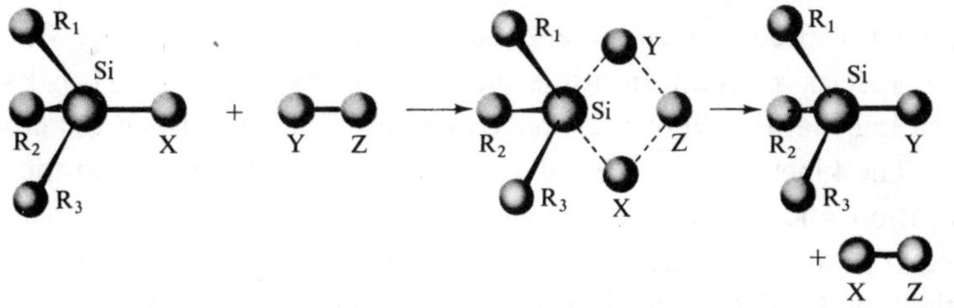

(b) X is a poor leaving group—substitution with retention as a result of adjacent attack. can assist the departure of X in a four-centre transition state

Fig. 4–6 Remote and adjacent attack at tetrahedral silicon

other hand, a poor leaving group, such as CH_3O or H, is replaced with retention of configuration and it is suggested that the entering and leaving groups are adjacent to one another in the transition state or intermediate because the leaving group requires some electrophilic assistance that it can gain from a cyclic transition state (Fig. 4-6). Other factors also play some part. For example, when the situation is otherwise fairly balanced a more polar solvent will tip the balance in favour of path (a) where the charge separation in the transition state is greater than in path (b).

The lower stringency of the stereochemical rules is seen in the substitution reactions at a bridgehead atom (Fig. 4-7). The three 'chelate rings' would effectively prevent bimolecular attack from the back of the reaction centre and the formation of a trigonal bipyramidal transition state or intermediate with the entering and leaving group in axial positions. If the

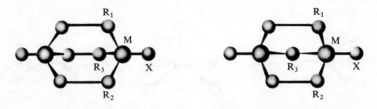

Fig. 4 – 7 Reaction centre at a bridgehead. Bimolecular attack from the rear is prevented. R_1, R_2, and R_3 are unable to become coplanar with the reaction centre in these examples

rings are small enough the three 'donor' atoms cannot become coplanar with the reaction centre. The reactivity of a bridgehead carbon is very greatly reduced whereas that of an analogous silicon atom can be reduced, enhanced, or unaffected depending on the nature of the reaction studied. In general, the availability of empty $3d$ orbitals not only makes an A mechanism more likely, it allows a wider range of arrangements of ligands in the five-coordinate intermediate.

4-5-2 Phosphorus

Phosphorus, as representative of group V can present a range of stable tetrahedral species which undergo substitution relatively slowly. Many of these species contain multiple bonds, for example, $(C_6H_5)_3P=O$, indicating that the $3d$ orbitals have been brought into use in the ground state. The kinetic studies reveal a strong dependence of rate on the concentration and nature of the entering group and therefore indicate an associative intimate mechanism. It has been suggested that anionic species of the type $Cl_2PO_2^-$ hydrolyse with a dissociative mechanism, but this is rare compared to the associatively activated processes.

The stereochemistry of substitution again depends upon the nature of the reaction. The rate of reaction

$$(C_2H_5)(C_6H_5)PO(OCH_3) + {}^*CH_3O^- \rightleftharpoons$$
$$(C_2H_5)(C_6H_5)PO(O^*CH_3) + CH_3O^-$$

(follow by studying the rate of distribution of ^{14}C-labelled methoxide) is exactly one half of the rate of racemization of the optically active substrate under identical conditions. This indicates that each act of substitution leads to a product with inverted configuration and thereby cancels out two units of optical activity. In some cases, complete retention is observed and in others the substitution is non-stereospecific. Many cases are to be found in cyclic phosphorus compounds where problems of ring-strain control the form of the five-coordinate intermediate and hence the steric course of substitution. For inversion of configuration the entering and leaving groups must take up axial positions in the five-coordinate trigonal bipyramid. This would leave a 120° bite for the 'chelate' now attached equatorially. A smaller ring might be less strained with the 90° resulting from the occupation of one axial and one equatorial position and only permit adjacent attack. With fairly electronegative entering and leaving groups, the tendency to take up axial positions may be opposed by ring strain considerations and it has been suggested that small rings will favour retention of configuration while larger rings will permit inversion (Fig. 4-8).

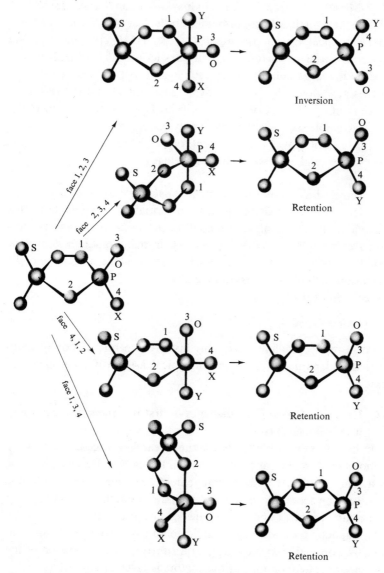

Fig. 4 – 8 Possible intermediates from nucleophilic attack on a phospholane. (A series of five other trigonal bipyramidal arrangements with Y in the trigonal plane can be formed by attack at five of the six tetrahedral edges. All will lead to retention of configuration)

4-6 Transition elements

4-6-1 Simple tetrahedral electrovalent complexes

These complexes have earned the reputation for being extremely labile and little work has been done on them. Nuclear magnetic resonance studies have shown that the rates of substitution are proportional to the concentration of free ligand, and associative mechanisms have been assigned to the reactions of $CoCl_4^{2-}$, $CoCl_3py^-$ (py = pyridine), $CoCl_2py_2$, $CoCl_2(2\text{-Mepy})_2$, (2-Mepy = 2-methylpyridine), and $CoCl_2(HMPA)_2$ (HMPA = hexamethylphosphoramide). In the last case, the kinetic form is more complicated, that is, rate = $(k_1 + k_2[HMPA])[\text{complex}]$, and the k_1 term is said to correspond to a parallel dissociative path that is assisted by the large bulk of the HMPA ligand. All of these tetrahedral complexes are potentially coordinatively unsaturated and the reactivity sequence $[CoCl_4]^{2-} \ll [CoCl_3py]^- \ll [CoCl_2py_2]$ is said to reflect the approach to stable six-coordination. It may be remembered that while the blue form of $[CoCl_2py_2]$ is tetrahedral, the violet form is an octahedral polymer in the solid state, with chloride bridges.

4-6-2 Tetrahedral oxy-anions and their simple derivatives

A considerable amount of mechanistic information is available in this area if one looks at the organic oxidation reactions of chromates, permanganates, vanadates, and so on because in many cases an important part of the reaction is the actual entry of the organic reagent into the coordination shell of the transition metal ion.

For a reaction of the type

$$L + O_3Cr-O-CrO_3^= \longrightarrow LCrO_3 + CrO_4^=$$

the rates of reaction, while reasonably fast are dependent upon the concentration and nature of L so that a typical associative mechanism is indicated.

4-6-3 High-spin tetrahedral complexes of first row transition metal ions with bulky ligands

It might have been possible to discuss these reactions together with those of Section 4-6-1 except for the fact that, in the case of nickel(II), there is often a four-coordinate planar arrangement available as either a stable alternative or an accessible intermediate. In the event, this does not appear to lead to any problem and the kinetics of ligand exchange followed by nuclear magnetic resonance again indicate an associative mechanism, in spite of the congestion in the tetrahedral substrate. An indication of the dependence of rate upon the various factors is given in Table 4-4.

The reactivity sequence Fe(II) > Ni(II) > Co(II) appears to hold throughout the range of reactions studied.

Table 4-4 Rate constants and activation parameters for tetrahedral substitution reactions of the type $MX_2L_2 + L^* \rightarrow MX_2LL^* + L$ in chloroform solution

M	X	L	$k_2(25°)M^{-1}s^{-1}$	$\Delta H^{\ddagger}$ kcal.mole^{-1}	$\Delta S^{\ddagger}$ cal.deg^{-1}mole^{-1}
Fe	Br	$P(C_6H_5)_3$	2.0×10^5	3.8	−22
Fe	Br	$P(p\text{-}CH_3.C_6H_4)_3$	6.3×10^5	4.0	−19
Co	Cl	$P(C_6H_5)_3$	1.2×10^4	8.8	−10
Co	Br	$P(C_6H_5)_3$	8.7×10^2	7.7	−19
Co	I	$P(C_6H_5)_3$	2.6×10^2	9.1	−17
Co	Cl	$P(p\text{-}CH_3.C_6H_4)_3$	2.2×10^3	4.4	−29
Co	Br	$P(p\text{-}CH_3.C_6H_4)_3$	1.8×10^3	5.5	−25
Co	I	$P(p\text{-}CH_3.C_6H_4)_3$	8.3×10^2	5.9	−25
Co	Br	$P(C_6H_5)_2(n\text{-}C_4H_9)$	8.3×10^3	7.1	−17
Co	I	$P(C_6H_5)_2(n\text{-}C_4H_9)$	1.1×10^2	7.7	−23
Ni	Cl	$P(C_6H_5)_3$	3.2×10^5	8.1	−7
Ni	Br	$P(C_6H_5)_3$	6.9×10^3	4.7	−25
Ni	I	$P(C_6H_5)_3$	6.4×10^2	6.9	−23
Ni	Cl	$P(p\text{-}CH_3C_6H_4)_3$	1.9×10^4	5.2	−21
Ni	Br	$P(p\text{-}CH_3C_6H_4)_3$	6.1×10^3	5.2	−24
Ni	I	$P(p\text{-}CH_3C_6H_4)_3$	5.9×10^3	4.4	−24

4-6-4 Covalent molecules obeying the '18 electron' rule

The d^{10} configuration requires four bonding pairs of electrons to make a valency shell that has the same electron configuration as the next noble gas. Such compounds are tetrahedral but now represent a situation where, although the four-coordinate system is open, no readily available orbitals are present to facilitate the associative mechanism. On the other hand, coordination unsaturation (three- and two- coordinate species) is known and becomes stable and common as the effective nuclear charge increases. Much work has been done in this area with $Ni(CO)_4$ and its derivatives. Provided care is taken to avoid facile exchange in the gas phase, it can be shown that the exchange of carbon monoxide and its replacement by other suitable ligands, such as phosphines and phosphites, takes place by way of a common intermediate and at a rate that is independent of the concentration and nature of the entering group. All the evidence points to a 'D' mechanism. The substitution reactions of $Ni(CO)_3L$ and $Ni(CO)_2L_2$ follow a similar pattern and the D mechanism is characteristic of this area. Although simple carbonyls of the congeners of nickel have not yet been described, compounds of the type $Pd(PPh_3)_4$ and $Pt(PPh_3)_4$ are well known and have been extensively studied. Although the interest here is mainly concerned with oxidative-addition (Chapter 10), kinetic studies have shown that the active species for the oxidation has a reduced coordination number and that a dissociative mechanism is therefore quite feasible.

In many of the book-keeping operations like the '18-electron' rule, the NO group can be counted as a three-electron donor and so a series of 'isoelectronic' tetrahedral compounds can be obtained by moving one place left in the periodic table for each CO replaced by NO. Thus it is possible to consider the tetrahedral species $Co(CO)_3NO$, $Fe(CO)_2(NO)_2$, $MnCO(NO)_3$, and $Cr(NO)_4$ as being isoelectronic with $Ni(CO)_4$. Extensive studies have been carried out on the substitution reactions of the first two members of this series. The characteristic feature of these reactions, which, incidentally, is common to many other compounds containing the NO group, is the sensitivity of rate to the concentration and nature of the entering nucleophile. Indeed, there is every reason to assign an associative mechanism to the substitution reactions of these complexes. It has been suggested that the NO group can assist the associative mechanism by changing its electron configuration. By considering NO as a three-electron donor we are effectively assuming that it coordinates as NO^+ under these circumstances, that is, $M \leftarrow N\equiv O$ (remember that NO^+ is isoelectronic with CO). If it is assumed that in the five-coordinate intermediate the problem of where to put the extra electrons is solved by changing the nitrosyl ligand from NO^+ to NO^-, the 18 electron rule is not violated. However, the process takes on the features of an oxidative addition and it does not require much stretching of the imagination to look upon the intermediate as a five-coordinate d^8 species (Fig. 4-9).

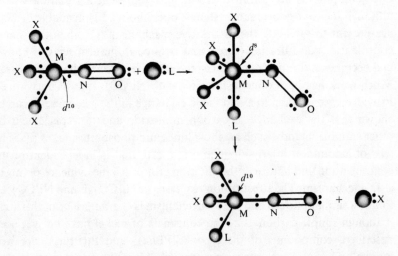

Fig. 4 – 9 Oxidative addition picture of bimolecular substitution at a formally d^{10} nitrosyl complex. An 18-electron shell is maintained about M. (Charges are omitted for purposes of clarity)

Problems

4-1 Sketch the possible transition states (or intermediates) for and discuss the factors that control the relative importance of adjacent and remote (front side and back side) attack in bimolecular nucleophilic substitution at tetrahedral silicon.

4-2 Sodium α-bromopropionate hydrolyses in aqueous solution to give sodium lactate. In the presence of hydroxide, the hydrolysis follows the rate law: rate = $(k_1 + k_2[\text{OH}^-])[\alpha\text{-bromopropionate}]$. An analysis of the product when $(+)_D$-α-bromopropionate is used indicates that the k_1 path yields lactate with $>85\%$ retention of configuration at carbon, whereas the k_2 path is accompanied by $>80\%$ inversion. Explain these observations in terms of the mechanism and steric course of substitution at tetrahedral carbon. [See Cowdrey, W. A., E. D. Hughes, and C. K. Ingold, *J. Chem. Soc.*, 1208 (1937).]

4-3 Why is it that the kinetics of CO exchange and substitution reactions of $Ni(CO)_4$ in solution are only reconcilable if the dead space above the reaction solution is kept as small as possible?

4-4 Substitution at tetrahedral carbon is generally very much slower than at an analogous silicon compound. An exception is found in the hydrolysis of $(C_6H_5)_3MCl$ in aqueous acetone where M = C is a more reactive compound than M = Si. Explain these observations and predict the effect of adding a base to each reagent.

Bibliography

Belloli, R., Resolution and stereochemistry of asymmetric silicon, germanium, tin, and lead compounds, *J. Chem. Ed.*, 1969, **46**, 640.

Sommer, L. H., *Stereochemistry, Mechanism, and Silicon*. McGraw-Hill, New York, 1965.

5 Substitution at four-coordinate planar reaction centres

5-1 Occurrence

The four-coordinate planar configuration is greatly restricted in its occurrence (Fig. 5-1). The three major areas are:

(i) Compounds of P-block elements with four bonding and two non-bonding pairs of electrons, for example, ICl_4^-, XeF_4.

(ii) Complexes of transition metal ions in which the ligands require a four-coordinate planar arrangement, for example, the bis dithiolene complexes:

$$\begin{bmatrix} NC-C=S \quad\quad S=C-CN \\ \diagdown\;M\;\diagup \\ NC-C=S \quad\quad S=C-CN \end{bmatrix}^{\pm n}$$

or complexes of the planar quadridentate phthalocyanin, porphyrin, corrin, and so on, ligands.

(iii) Complexes of metal ions with a d^4, d^9, and low spin d^8 configuration.

Of these three categories, the first has not been studied in any depth and the second is irrelevant in that replacement of the in-plane ligands is either complicated or else destroys the planarity of the complex.

The d^4 and d^9 configurations generally have weak but significant bonding in the two remaining axial positions which represent a point of sensitivity for substitution. The reactivity of d^4 and d^9 metal ions of the first row of the transition series is very high indeed. Cr_{aq}^{2+} and Cu_{aq}^{2+} undergo substitution reactions with half-lives of approximately 10^{-8} seconds at 25°. They are possibly better looked upon as octahedral species with a long tetragonal distortion because the change from axial to equatorial coordination can be achieved by a single vibration mode of the complex.

For these reasons, any discussion of substitution in four-coordinate planar complexes must, at present, be limited to compounds where the central metal atom has the d^8 configuration. This is a common electron configuration and the known occurrences are listed in Table 5-1.

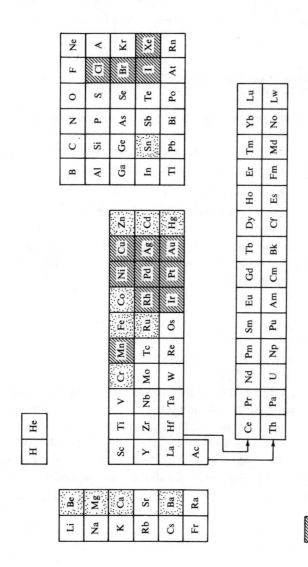

Fig. 5 – 1 Occurrence of planar four coordination

Table 5-1 Known oxidation states with the d^8 configuration

Cr(−II)	Mn(−I)	Fe(0)	Co(I)	Ni(II)	Cu(III)
Mo(−II)	Tc(−I)	Ru(0)	Rh(I)	Pd(II)	Ag(III)
W(−II)	Re(−I)	Os(0)	Ir(I)	Pt(II)	Au(III)

▓ Square planar complexes known and studied kinetically

5-2 Coordination number and geometry associated with the d⁸ configuration

A major feature in any discussion of the chemistry of the d^8 configuration is the interplay between four-, five-, and six-coordination. Apart from Ni(II) and Cu(III) one need consider only the low spin configuration. Ni(II) presents a wealth of possibilities and includes:

(i) high spin regular octahedral;
(ii) high and low spin tetragonally distorted octahedral;
(iii) trigonal prismatic six-coordinate;
(iv) high and low spin five-coordinate, with geometries ranging from regular trigonal bipyramidal to regular square pyramidal;
(v) high spin tetrahedral four-coordinate (discussed in Section 4-1);
(vi) low spin four-coordinate planar complexes.

The accessibility of alternative forms, often energetically close to the ground state, allows a large number of facile paths for reaction and makes any systematic study of the substitution reactions of square planar nickel(II) complexes rather difficult.

The oxidation state (+III) (with the exception of Au(III)) is unstable with respect to reduction and oxidation states (−II), (−I), and (0) are invariably either five-coordinate or else involved in cluster formation (with or without homonuclear bonds). The (+I) oxidation state is readily oxidized and here d^8 (four-coordinate) $\rightleftharpoons d^6$ (six-coordinate) changes become extremely important. For this reason the studies of substitution in four-coordinate planar complexes are restricted to Rh(I), Ir(I), Ni(II), Pd(II), Pt(II), and Au(III). The area is completely dominated by Pt(II) where factors such as reactivity, redox stability, and choice of reaction all achieve their maximum assistance to the research worker.

5-3 Quasi-theoretical arguments

The common occurrence of stable five-coordination in d^8 systems suggests strongly that the four-coordinate complexes are coordinatively 'unsaturated' and that a mechanism in which the central atom experiences a temporary increase in coordination number will be favoured for the substitution reactions. The formation of five equivalent bonds leads to an 18-electron valence shell (10 bonding and eight non-bonding electrons) which is the electronic configuration of the next noble gas. The utilization of all five orbitals $[2 \times (n-1)d + ns + 3 \times np]$ is favoured by a low effective nuclear charge.

It is still possible to find many examples of stable five- and six-coordinate complexes of low spin Pd(II), Pt(II), and Au(III) and two types of compound can be recognized:

(i) the essentially covalent species with five equivalent bonds, for example, $[Pt(SnCl_3)_5]^{3-}$ which closely resemble the typical compounds characterizing the (−II), (−I), and (0) oxidation states;

(ii) that with four strong bonds in a plane and one or two ligands attached in axial positions. These ligands are bonded relatively weakly by essentially electrostatic forces and the bonds are unduly long. A typical example is $[Au(diars)_2I_2]^+$ (diars = o-phenylenebisdimethylarsine). Here the high effective nuclear charge leads to a strong electrostatic attraction for the two anionic ligands.

All of these considerations point towards an associative mechanism as the favoured pathway for substitution and it ought to be possible to relate what we know about stable five-coordination in these systems to the energetic relationships between the labile five-coordinate intermediates and transition states and the four-coordinate ground state. In this way it becomes possible to understand the factors controlling reactivity.

5-4 Direct kinetic evidence

In those reactions where the relationship between rate and concentration can be studied, a common rate law is observed. For a reaction represented by

$$[R_3MX] + Y \longrightarrow [R_3MY] + X$$

this takes the form

$$\frac{-d[R_3MX]}{dt} = (k_1 + k_2[Y])[R_3MX]$$

For a particular substrate in a given solvent at a fixed temperature, k_1 is independent of the nature of Y, whereas k_2 is very sensitive to the nature of Y. Occasionally k_1 is too small to be observed and this is quite a common feature in substitution reactions in non-coordinating solvents or in many of the reactions of Au(III) complexes. In other cases the k_2 term vanishes and the rate is independent of the concentration of Y. This situation is sometimes, but not usually, indicative of a change in mechanism and will be discussed later.

The k_2 path has all the characteristics of an associative intimate mechanism. Direct kinetic evidence for an intermediate, in terms of departures from the simple rate law, is rather rare and occasionally ambiguous, but the fact that it is often possible to consider the bond-making aspects of substitution independently of the bond-breaking aspects suggests most strongly that a short-lived five-coordinate intermediate is involved.

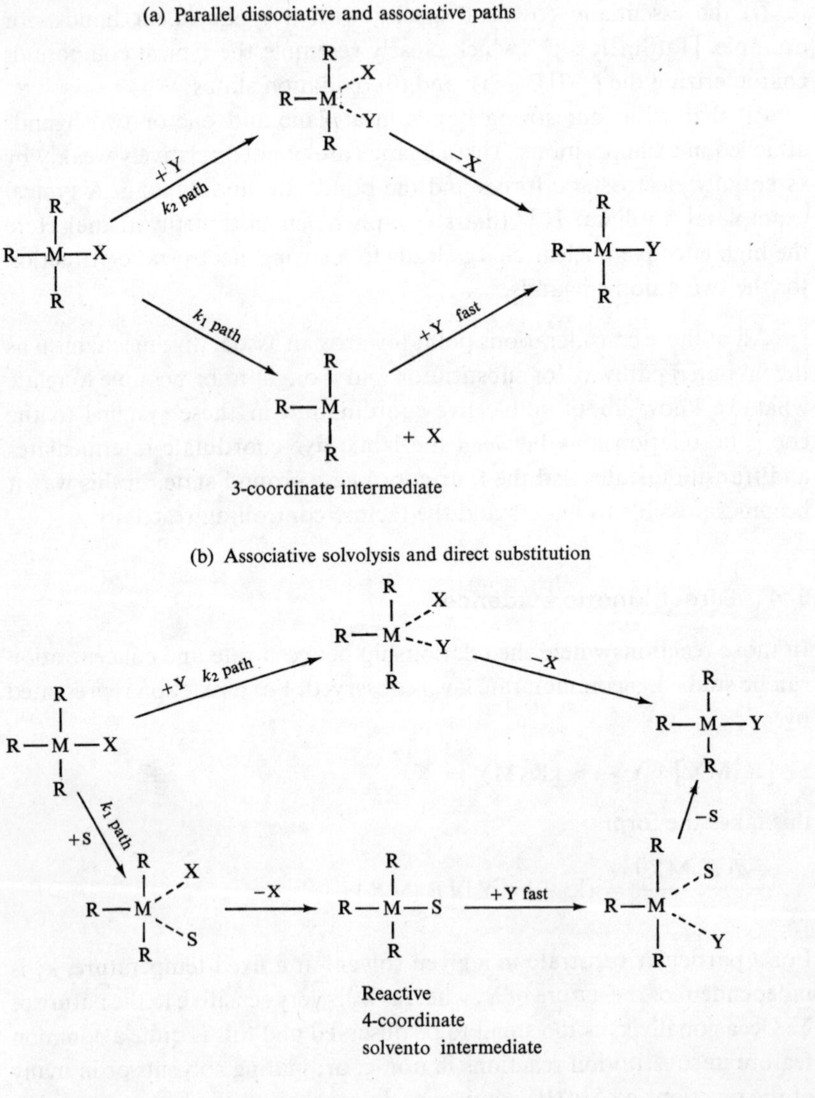

Fig. 5-2 Two limiting mechanisms to account for the nucleophile independent path of four-coordinate planar substitution

Sec. 5-4   Direct kinetic evidence

The major mechanistic problem in square planar substitution is the assignment of the k_1 term. It is obvious that the nucleophile Y is not involved in the rate-determining transition state, but what role does the solvent play in the process? The two limiting possibilities are shown in Fig. 5-2. Both possible k_1 paths require a reactive intermediate, which in the dissociative path will be the three-coordinate R_3M species and in the associative path will be the solvento complex R_3MS. Direct kinetic distinction, by competitive reactions or mass-law retardation (reduction of rate by increasing the concentration of the leaving group, X), will not differentiate the two types of reactive intermediate.

However, the following independent arguments allow a distinction to be made:

(i) The repeated appearance of the two-term rate law requires that the factors controlling the energetics of the k_1 and k_2 path are similar. Bond-weakening substituents do not favour the k_1 path and suppress the k_2 path, neither do bond-strengthening substituents function in the opposite way. This would suggest that both paths have the same mechanism.

(ii) Steric congestion in the reagent generally retards both the k_1 and the k_2 path, suggesting that both have an associative mechanism.

(iii) The k_1 term is very sensitive to the coordinating power of the solvent and vanishes in non-coordinating solvents. For example, the reaction

$$\textit{trans-}[PtCl_2(PEt_3)(NHEt_2)] + \text{amine} \longrightarrow$$
$$\textit{trans-}[PtCl_2(PEt_3)\text{amine}] + NHEt_2$$

has been studied by Raethel and Odell and is of interest since all reagents and products are uncharged and the reaction can be followed in a very wide range of coordinating and non-coordinating solvents. In hexane or benzene the k_1 term vanishes and rate = k_2[complex][amine], while in methanol the kinetics take on the usual form

$$\text{rate} = (k_1 + k_2[\text{amine}])[\text{complex}].$$

The complex is extremely crowded and the steric hindrance reduces the values of both k_1 and k_2 by factors of about 10^3 compared to the unhindered system. It is of interest to note that many reactions with reasonably large amines go entirely by way of the k_1 path, indicating quite clearly that the congestion produced by two bulky amines attached to the platinum at the same time is energetically too unfavourable. All of these features are consistent with the interpretation of the k_1 path in terms of an associative solvolysis.

(iv) Olcott and Gray have been able to trap a solvent containing intermediate and hence demonstrate an associative mechanism for the k_1

term in the reactions of [Pt dien I]$^+$ (dien = diethylenetriamine) in water. The solvento complex [Pt dien H$_2$O]$^{2+}$ can be prepared independently and shown to be extremely labile with respect to the displacement of the coordinated water. The removal of a proton, however, is faster than any process requiring the fission of the Pt—O bond and the resultant [Pt dien OH]$^+$ complex is inert to substitution. Fortunately, the base OH$^-$ is a very weak nucleophile in the Pt(II) system and does not generate a k_2 term. It is therefore possible to trap any solvento complex formed by the k_1 path before the water can be replaced by Y. The process that was followed was the formation of [Pt dien OH]$^+$ from [Pt dien I]$^+$ in the presence of varying amounts of OH$^-$ and I$^-$. If the k_1 path is dissociative:

$$[\text{Pt dien I}]^+ \underset{k_{-1}}{\overset{k_1}{\rightleftharpoons}} \text{Pt dien}^{2+} + \text{I}^-$$

$$\text{Pt dien}^{2+} + \text{H}_2\text{O} \xrightarrow{k_2} \text{Pt dien H}_2\text{O}^{2+}$$

$$+\text{OH}^- \updownarrow +\text{H}^+ \quad \text{very fast}$$

$$\text{Pt dien}^{2+} + \text{OH}^- \xrightarrow{k_3} \text{Pt dien OH}^+$$

then the rate of formation of [Pt dien OH]$^+$ should be retarded by iodide and possibly accelerated (within the limit of the slow dissociation) by hydroxide. However, if the k_1 path arose from an associative solvolysis:

$$[\text{Pt dien I}]^+ + \text{H}_2\text{O} \xrightarrow{k_1} [\text{Pt dien OH}_2]^{2+} + \text{I}^- \quad \text{associative}$$

$$[\text{Pt dien OH}_2]^{2+} + \text{OH}^- \longrightarrow [\text{Pt dien OH}]^+ + \text{H}_2\text{O} \quad \text{very fast}$$

the rate of formation of [Pt dien OH]$^+$ would be independent of the concentrations of I$^-$ and OH$^-$, provided there was enough of the latter to deprotonate all the aquo complex that was formed. The latter was proved to be the case and so, in this reaction at least, the k_1 term has been identified with a bimolecular solvolysis.

The evidence is therefore fully in favour of an associative mechanism for both substitution paths and this is the normal mode of reaction in a low spin d^8 four-coordinate complex. In very special circumstances a dissociative mechanism can be reluctantly wrung from the complex by suppressing the alternative reaction paths. This occurrence is sufficiently rare at the moment to warrant a separate discussion.

5-5 Geometry of the transition states and intermediates

Since the standard mechanism for substitution in these four-coordinate planar complexes is associative the intermediates and transition states will be (at least) five-coordinate. This geometry can be characterized by two regular figures (and all possible distortions between them) but the bulk of

Sec. 5–5   Geometry of transition states and intermediates   49

the evidence supports the trigonal bipyramid with the entering and leaving groups in the trigonal plane (together with the *trans* ligand) and the two ex-*cis* ligands in axial positions. This is true for the intermediate (where the entering and leaving groups will be equivalently bonded) and the transition states, where there will be disparity between the forming and breaking bonds. The square pyramidal arrangement with the incoming group in the apical position may very well be passed along the reaction coordinate but normally this would be immaterial to the kinetics of the process unless a

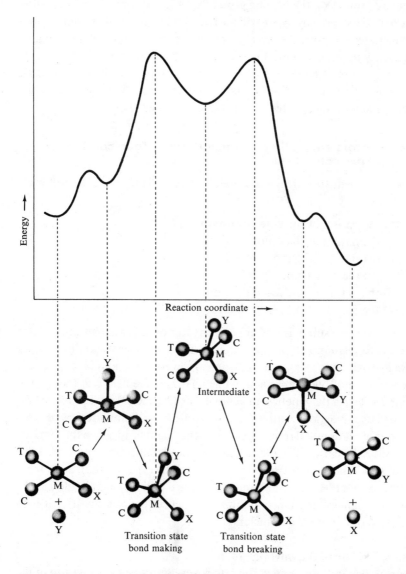

Fig. 5 – 3 Diagrammatic reaction profile for an A-type substitution in a planar four-coordinate complex

significant fraction of the substrate was associated with Y in a pre-equilibrium. Involvement of solvent molecules above and below the plane is implied but, as was pointed out in Section 3-2, solvation changes on activation should not be mixed up with molecularity. These relationships are all shown in Fig. 5-3 where it will be seen that, under normal circumstances, substitution must take place with complete retention of configuration and this is indeed what is found. Exceptions to this rule are of considerable interest. The relationship between the positions of the entering group (Y), the leaving group (X), and the *trans* ligand (T), allow one to understand why, as we will see later, many ligands that exert a strong labilizing effect from the *trans* position are also very effective reagents when they function as entering group. Another feature that requires considerably more discussion is the relative importance of the bond-making and bond-breaking transition states and the extent to which they are independent or interrelated. This will be examined later.

5-6 Factors controlling the reactivity of four-coordinate planar complexes

It is convenient to split the discussion in accordance with the following headings:

(1) the nature of the entering group;
(2) the nature of the other ligands:
   (a) the *trans* ligand;
   (b) the *cis* ligand.
(3) the nature of the leaving group;
(4) the nature of the reaction centre.

The various subdivisions are to a greater or lesser extent interrelated and the discussion will be, from time to time, somewhat akin to juggling with five balls in the air at the same time. However, it is possible to consider the variation of one component while all others are held constant and to see the effect on the kinetics. In general, (1), (2), and (3) are considered for a particular reaction centre and the data are so dominated by the behaviour of platinum(II) complexes, that one might be forgiven for thinking that this is the pattern for all four-coordinate planar substitution. That this is not so will be demonstrated in (4). The effects (1), (2a), (2b), and (3) are arranged in order of decreasing dominance and a general rule can be applied—namely, the more dominant the effect the less sensitive it is to variation of the other parameters.

5-6-1 Nature of the entering group

In an associative mechanism the rate is very sensitive to the nature of the entering group. The effectiveness of the entering group can be measured by the magnitude of the rate constant k_2 and, for substitution reactions involving Lewis bases as entering and leaving groups, it is usual to use the term **nucleophilicity** for this quantitative measure of reactivity. One of the simplifying features of substitution reactions of platinum(II) complexes is that it is possible to set up a quantitative scale of nucleophilicity and to show that it holds for a wide range of substrates and reactions and therefore can be used predictively. The development of nucleophilicity scales did not start with platinum(II). Many years ago Swain and Scott set up a quantitative scale for tetrahedral carbon, and interest in the relationship between the factors controlling nucleophilicity and the nature of the reaction centre was one of the features that led Pearson to develop the concept of 'hard and soft acids and bases'. To set up an empirical scale of nucleophilicity it is necessary to choose a standard reaction. For platinum(II), the reaction chosen was

$$trans\text{-}[Pt\ py_2Cl_2] + Y^- \longrightarrow trans\text{-}[Pt\ py_2Cl\ Y] + Cl^-$$

in methanol at 30°. This follows the usual rate law:

$$\text{rate} = (k_1 + k_2[Y^-])[Pt\ py_2Cl_2],$$

and the nucleophilicity of Y, written as $n_{Pt}(Y)$ is defined by

$$n_{Pt} = \log(k_2/k_1).$$

Since k_2 (litres.mol^{-1} s^{-1}) and k_1(s^{-1}) have different dimensions and k_1 really represents bimolecular attack by solvent methanol, it was thought

Table 5-2 A selection of n_{Pt}^o values for a range of nucleophiles

Nucleophile	Donor	n_{Pt}^o	Nucleophile	Donor	n_{Pt}^o	Nucleophile	Donor	n_{Pt}^o
CH_3OH	O	0·00	$(C_6H_5CH_2)_2S$	S	3·29	$C_6H_{11}NC$	C	6·20
CH_3O^-	O	<2·4	C_6H_5SH	S	4·15	CN^-	C	7·00
			$(C_2H_5)_2S$	S	4·38			
NH_3	N	3·06	$(CH_3)_2S$	S	4·73	$(CH_3O)_3P$	P	7·08
NC_5H_5	N	3·13	SCN^-	S	5·65	$(C_6H_5)_3P$	P	8·79
NO_2^-	N	3·22	SO_3^{2-}	S	5·79	$(C_4H_9)_3P$	P	8·82
N_3^-	N	3·58	$C_6H_5S^-$	S	7·17	$(C_2H_5)_3P$	P	8·85
NH_2OH	N	3·85	$SC(NH_2)_2$		7·17			
NH_2NH_2	N	3·85	$S_2O_3^{2-}$	S	7·34	$(C_6H_5)_3As$	As	6·75
						$(C_2H_5)_3As$	As	7·54
F^-	F	<2·4	$(C_6H_5CH_2)_2Se$	Se	5·39			
Cl^-	Cl	3·04	$(CH_3)_2Se$	Se	5·56	$(C_6H_5)_3Sb$	Sb	6·65
Br^-	Br	3·98	$SeCN^-$	Se	7·10			
I^-	I	5·42						

desirable to convert k_1 to a second-order constant by means of the relationship $k_1/[CH_3OH] = k_1^\circ$ (litres.mol^{-1} s^{-1}). This produces a dimensionless n_{Pt}° defined by $n_{Pt}^\circ = k_2/k_1^\circ = k_2[MeOH]/k_1$. The concentration of methanol in pure methanol at 30° is 24·9 M and so $n_{Pt}^\circ = n_{Pt} + 1\cdot 40$ ($1\cdot 40 = \log 24\cdot 9$). A selection of n_{Pt}° values is given in Table 5-2.

A number of important features can be seen from studying this list and combining the information with that gathered from other sources:

(i) Hydroxide, which is often a very good nucleophile towards other reaction centres, is completely ineffective and never seems to produce a k_2 term. A satisfactory explanation of the *complete lack* of bimolecular behaviour has yet to be provided.

(ii) The halides increase their nucleophilicity along the sequence

$$F^- \ll Cl^- < Br^- < I^-.$$

(iii) The 'light element' donors, N, O, F, are less effective than their congeners, thus

$$R_3P > R_3As \gg R_3N$$
$$R_2S \gg R_2O$$

(iv) The nucleophilicity is not greatly sensitive to proton basicity. Thus for a series of amines whose basicity covers a wide range (pK$_a$ range 4 → 10) the n_{Pt}° values cover less than one unit.

In general, the factors that promote nucleophilicity at platinum(II) are those which we associate with class (b) or 'soft' behaviour, which is not surprising in view of the fact that the reaction centre is also typically class (b) or 'soft'. This is characterized by the 'micropolarizability' of the donor rather than its proton basicity. Attempts to correlate n_{Pt}° values with any non-kinetic property of the ligand have met with little or no success. There is obviously no point in setting up a nucleophilicity scale if it only applies to the standard reaction and, fortunately, the reactivity sequence can be applied in quantitative terms to a wide range of platinum(II) complexes. In general, it is possible to find a linear relationship between $\log k_2$ and the n_{Pt}° value of the nucleophile concerned. Thus, for many reactions of the type

$$[R_3PtX] + Y \xrightarrow{k_Y} [R_3PtY] + X$$
$$\log k_Y = S \cdot n_{Pt}^\circ + c$$

where k_Y is the second-order rate constant for attack by Y, and n_{Pt}° is the appropriate nucleophilic reactivity factor for Y. The slope S, is termed the **nucleophilic discrimination factor** since the larger the slope, the more sensitive is the rate to the nature of the entering nucleophile. The intercept has been termed the **intrinsic reactivity** and while less easy to relate to reality than S, it is possible to say that $C = \log k_1 - 1.40$ where k_1 would be the first-order rate constant for the solvolytic path were the reactions to be carried out in methanol at 30°. As the properties of the actual solvent depart significantly from those of methanol, the reality of C vanishes. Fig. 5-4 shows the relationship between $\log k_2$ and n^o_{Pt} for a number of reactions that follow the linear relationship.

A very small number of nucleophiles do not appear to obey this simple linear relationship and change their nucleophilicity in response to the properties of the substrate. Cattalini has listed NO_2^-, $SC(NH_2)_2$, and $SeCN^-$ and a knowledge of general chemical behaviour would suggest that the reactivity of nucleophiles such as CO and olefins is very much dependent upon the nature of the substrate. It is usually found that when the effective nuclear charge from the platinum is rather high, or the complex is a little less 'soft' or class-(b)-like, the anomalous reagents are less effective than might have been predicted from their n^o_{Pt} values. Examples can be found in $[Pt\ dien\ H_2O]^{2+}$ or $[Pt\ dienCl]^+$. On the other hand, when the effective nuclear charge from the platinum is low, as in $[Pt\ Cl_4]^{2-}$, these nucleophiles are considerably more reactive than might be predicted.

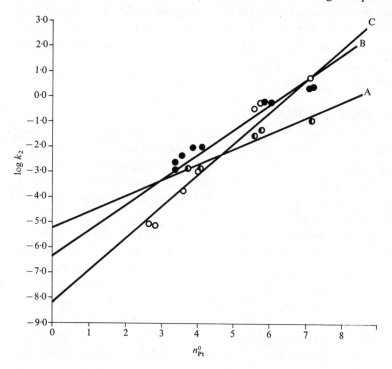

Fig. 5–4 Log k_2 vs n^o_{Pt} for the reactions of A, ◐, PtenCl$_2$ (S = 0·62, C = 5·25) B, ●, *trans*-[Pt(pip)$_2$Cl$_2$] (S = 1·00, C = −6·30) C, ○, *trans*-[Pt(PEt$_3$)$_2$Cl$_2$] (S = 1·26, C = −8·15)

Cattalini has suggested that these reagents are exhibiting **biphilic** properties in the sense that they are able to offer vacant orbitals to accept charge from the 'non-bonding' platinum electrons. This electrophilic π-contribution will depend upon the availability of the appropriate d orbitals of the metal and it was presumed that this availability would increase in the order [Pt dien H_2O]$^{2+}$ < [Pt dien Cl]$^+$ < trans-[Pt py$_2$ Cl$_2$] < [Pt Cl$_4$]$^{2-}$.

A second source of disturbance of the nucleophilicity sequence arises in sterically hindered systems where the size and shape of the nucleophile becomes more important than its n_{Pt}^o value. This is found in the reactions of trans-[PtCl$_2$(n-Pr$_3$P)(NHEt$_2$)] with amines and in the reaction of [Pd-(Et$_4$dien)SeCN]$^+$(Et$_4$dien = 3,9-diethyl-3,6,9-triazaundecane) with Br$^-$ and I$^-$ where the usual nucleophilicity is reversed and the smaller bromide ion is more reactive than the iodide.

Once we move away from the complexes of platinum(II) the quantity of systematic data rapidly decreases. The greater reactivity of palladium(II) and nickel(II) complexes restricts the range of reagents that can be conveniently studied. Nevertheless people are working on these systems in the hope of extending the field. Gold(III) complexes can be easily reduced to the d^{10} Au(I) and so the range of nucleophiles that can be examined is restricted but even with such restricted data it is clear that the n_{Pt}^o values cannot be applied quantitatively to reactions of gold(III) complexes and, perhaps more important, an n_{Au}^o scale with respect to a standard gold(III) substrate will be of little value because the reactivity order often depends upon the nature of the substrate. This interdependence of nucleophilicity and the nature of the substrate is an important feature of Au(III) chemistry.

5-6-2 Effect of the other ligands in the complex

In a four-coordinate planar geometry, once the leaving group has been identified, the remaining three ligands are separated by symmetry into a pair of ligands *cis* to the leaving group and one ligand *trans* to it. In the trigonal bipyramidal transition states these ligands retain their symmetry inequivalence and so it is meaningful to separate the discussion into *trans* effects and *cis* effects.

(a) *Effect of the* trans *ligand.* The *trans* effect has been the subject of much discussion in preparative, structural, thermodynamic, and kinetic studies. It takes its place among the legendary beasts of modern inorganic chemistry such as the principle of electroneutrality, the chelate effect, electronegativity, and so on, but unlike some of its illustrious bedfellows it has become, in recent years, amenable to quantitative study. Initially the postulation of the *trans* effect was the result of experience gained in preparative chemistry where it was found that some ligands were more able than others to promote substitution in a position *trans* to themselves in square planar platinum(II) complexes and that it was possible to set up an order of effectiveness which was maintained irrespective of the nature of the reaction concerned. This was of considerable importance in the design of specific synthetic paths and helped to boost the ego of the inorganic chemist who looked upon the organic chemist's planned approaches to synthesis with something resembling envy. It took approximately 20 years before people were able to ask, let alone answer, the right sort of question relating to the *trans* effect and it then became clear that the phenomenon was kinetic rather than thermodynamic (see Section 1–4) and hence must be related to the mechanism of the reaction. Kinetic studies indicated clearly that this is a *trans* labilizing effect and not a *cis* stabilizing effect. The best definition to date, of the *trans* effect is 'the effect of a coordinated ligand on the rate of replacement of a ligand *trans* to itself.' On this basis, a sequence of *trans* effectiveness can be assembled:

$H_2O \sim HO^- \sim NH_3 \sim$ amines $< Cl^- \sim Br^- < SCN^- \sim I^- \sim$
$NO_2^- \sim C_6H_5^- < CH_3^- \sim SC(NH_2)_2 <$ phosphines $\sim$ arsines $\sim$
$H^- \sim$ thioethers $<$ olefins $\sim CO \sim CN^-$

While the data do not exist for a quantitative comparison of a series as wide as this, it is possible to say that the increase in reactivity in going from T = Cl to T = H in the reaction of *trans*-[Pt(PEt$_3$)$_2$TCl] is about 10^5-fold.

A great deal of deduction has been based on the kinetics of reaction of these complexes with pyridine. Unfortunately, pyridine is such a weak nucleophile that a number of interesting effects have been missed. More recent work has shown that the *trans* effect sequence depends upon the nature of the nucleophile, (Table 5-3). In fact, the nucleophilic discrimination factor for *trans*-[Pt(PEt$_3$)$_2$RCl] increases along the sequence R = CH$_3$(0·74) < Ph(0·84) < Cl(1·26) whereas the intrinsic reactivity decreases $-4·13 > -5·27 > -8·15$ respectively. The reactivity towards

Table 5-3 Second-order rate constants ($M^{-1}s^{-1}$) for the reaction of nucleophiles with *trans*-[Pt(PEt$_3$)$_2$RCl] in methanol at 30°

Nucleophile	R = CH$_3$	R = C$_6$H$_5$	R = o-tolyl	R = Cl
NO$_2$	1·6	0·45	0·0[a]	0·0027
N$_3^-$	7·0	0·8	0·0[a]	0·02
Br$^-$	11·6	1·8	0·0[a]	0·093
I$^-$	40	6·0	0·0[a]	23·6
thiourea	15,000	630	65·2	very fast
CN$^-$		3,610	23·4	

[a] Rate independent of concentration of nucleophile.

weak nucleophiles is determined mainly by the intrinsic reactivity and appears to follow the classical *trans* effect sequence whereas the reactivity towards strong nucleophiles tends to reflect the nucleophilic discrimination factor. Therefore it is dangerous to take too detailed a view of the quantitative aspects of the *trans* effect before adequate data are available. In recent years the term has also been applied to a range of phenomena of a non-kinetic and non-substitutional character. For example, it has been related to bond lengths, force constants, and stretching frequencies.

In discussing the effect it is convenient to separate it into two parts:

(i) ground state effects;
(ii) transition state effects.

The ground state effects can be studied by non-kinetic methods and can be observed in equilibria, infrared, n.m.r., and e.s.r. spectroscopy and in the variation in bond length. For a series of compounds of the type *trans*-$[ML_2T.X]$, where M, L, and X are held reasonably constant and T is varied, it is assumed that the M—X bond length is related to the strength of the bonding—the longer the bond, the weaker the link. A considerable amount of accurate data are now available to show that the length of the Pt—Cl bond is quite sensitive to the electronegativity of the ligand in the *trans* position—the less electronegative the ligand, the longer the PtCl bond. Some values are listed in Table 5-4. It is of interest to note that $\diagup\!\!\!\diagdown C=C\diagup\!\!\!\diagdown$ does not have a strong ground state *trans* effect. The effect of bond weakening leads to a destabilization of the ground state and would invariably lead to an increase in reactivity in a dissociative process.

Table 5-4 Ground state weakening *trans* effect. Effect of the electronegativity of the *trans* ligand upon the Pt—Cl bond length

Compound	Trans *ligand* donor atom	Electro-[b] negativity of donor	Pt—Cl Å
$(Pt(acac)_2Cl]^{-}$ [a]	O	3·50	2·28
$[PtCl_3NH_3]^{-}$	N	3·05	2·32
trans-$[Pt(PEt_3)_2Cl_2]$	Cl	2·83	2·30
$[(C_{12}H_{17})_2Pt_2Cl_2]$	C=C	2·75	2·31
cis-$[Pt(PMe_3)_2Cl_2]$	P	2·15	2·37
trans-$[Pt(PPh_2Et)_2HCl]$	H	2·1	2·42
trans-$[Pt(PPhMe_2)_2(SiPh_2Me)Cl]$	Si	1·90	2·45

[a] acac = acetylacetonate
[b] Pauling scale
  Trans effect sequence Si > H > P > C=C ~ Cl > O

In an associative process, on the other hand, matters are more finely balanced and it is pertinent to ask whether this bond weakening destabilizes the ground state more or less than it does the transition state. Langford and Gray have pointed out that, of all the orbitals involved in the bonding, only the p orbitals specifically involve *trans* pairs of atoms. If a pair of atoms of widely disparate electronegativity are sharing the same p-orbital then the less electronegative atom will develop its covalent bond at the expense of the more electronegative one. In the trigonal bipyramidal transition state this competition is eased, since the pair of atoms no longer compete for the same orbital in such a direct fashion. This can be seen in Fig. 5-5. The ground state *trans* effect is essentially an inductive or σ effect and the electronegativity of the donor atom will of course be modified by the groups that are attached to it.

The transition state *trans* effect can only be seen in the rate of the reaction and arises from a relative stabilization of the transition state. This was at one stage believed to be the major cause of the *trans* effect. Chatt, and later Orgel, were the first to notice that many strong *trans* effect ligands had the ability (but not necessarily the inclination) to accept charge from the metal into empty antibonding or non-bonding orbitals and thereby reduce the accumulation of charge resulting from the attachment of the incoming nucleophile (Fig. 5-6).

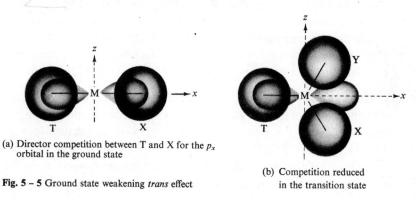

(a) Director competition between T and X for the p_x orbital in the ground state

(b) Competition reduced in the transition state

Fig. 5–5 Ground state weakening *trans* effect

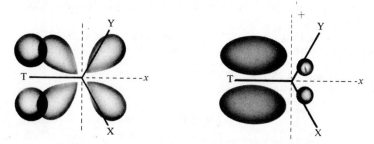

Fig. 5–6 Presence of an empty orbital of π-symmetry on T will allow withdrawal of charge in d_{xz} orbital away from X and Y towards T. This stabilizes the five-coordinate species

This effect, if operating in the ground state, can actually strengthen the bonding of the other ligand unless it too is competing for π-bonding. It has been calculated that the transition state, or π-*trans* effect, follows the order

$$\text{>C=C<} \ \ CO > CN^- > NO_2^- > -SCN^- > I^- > Br^- > Cl^-$$
$$> NH_3 > OH^-$$

The actual nature of the ligand is now of more importance than the nature of the donor, since a conjugative effect will depend upon the whole group of atoms over which the electrons are delocalized.

(b) *Effect of the* cis *ligand*. Compared with the *trans* effect, this is rather weak and, being weak, depends upon a large number of variables. It is possible at the moment to detect two types of behaviour which appear to be significant, but no guarantee can be given that this significance will remain as more and more data are collected.

(i) In the presence of a strong *trans* effect ligand. It appears that donors follow the same labilizing sequence when they are *cis* to the leaving group as when they are *trans*. Thus, in the complex *cis*-[Pt(PEt$_3$)$_2$RCl] the rate of replacement of Cl under the strong *trans* influence of PEt$_3$ increases along the sequence R = Cl < Ph < CH$_3$. The k_1 values obtained from the reaction with pyridine are given in Table 5-5. The reactivity is dominated by the *trans* labilizing ligand and, since the whole range of reactivity is covered by a factor of three, it may well be fortuitous that the above mentioned sequence was observed at all.

(ii) In the presence of a weak *trans* effect ligand. When the *trans* ligand and the nucleophile are weak the *cis* effect tends to dominate the reactivity. Thus the rate constants for the various substitutions in $[M(NH_3)_n(Cl)_{4-n}]^{(n-2)+}$ (M = Pd or Pt, $n = 0 \rightarrow 4$) can only be rationalized if the *trans* effects of NH$_3$ and Cl are considered to be virtually identical but the *cis* effect of NH$_3$ is considerably greater than

Table 5-5 Values of $k_1(s^{-1})$ for the reaction of *cis* and *trans*-[Pt(PEt$_3$)$_2$RCl] with pyridine in ethanol

cis ligands = PEt$_3$, PEt$_3$ *trans* ligand, R, varied		*cis* ligand, PEt$_3$, *trans* ligand PEt$_3$ *cis* ligand, R, varied	
R	$10^4 k_1 (s^{-1})(25°)$	R	$10^4 k_1 (s^{-1})(0°)$
H	180		
CH$_3$	1·7	CH$_3$	600
C$_6$H$_5$	0·33	C$_6$H$_5$	380
Cl	0·010	Cl	170

that of Cl. This is especially true for the Pd(II) complexes. The same appears to be true when trans-$[Pt(PEt_3)_2Cl_2]$ is compared with *trans*-$[Pt(py)_2Cl_2]$. As far as the reaction with a weak nucleophile like Cl^- or NO_2^- is concerned the *cis* PEt_3 ligands are strongly deactivating ($k_{NO_2^-} = 2 \cdot 7 \times 10^{-5} M^{-1} s^{-1}$, $6 \cdot 8 \times 10^{-4} M^{-1} s^{-1}$ at 30° in methanol for the PEt_3 and py complexes respectively). However, with strong nucleophiles the *cis* discrimination appears to be far less important and the reactivity is very much dominated by the strong nucleophile for example, $k_{SCN^-} = 0 \cdot 371$ and $0 \cdot 180$ $M^{-1} s^{-1}$ for the phosphine and pyridine complexes respectively in methanol at 30°). The reason for this is that the nucleophilic discrimination factors are affected by the *cis* ligands (see Fig. 5-4).

5-6-3 Nature of the leaving group

This is by far the most difficult effect to systematize because it is intimately connected with the nature of the entering nucleophile and the *trans* ligand. As a rule, the study must be limited to the most replaceable ligand in the complex and the leaving ability of a particular ligand must be a combination of the intrinsic lability of the ligand itself (if such a concept has any real meaning) and the labilizing effects exerted upon it by the other ligands in the complex. Unless one is prepared to work with a very strong *trans* labilizing ligand, leaving-group variation must be restricted to the less strongly bonded ligands. In a dissociative mechanism the lability will depend upon the factors that promote bond dissociation (for example, bond strength, electron displacement effects, and so on) and the compensating solvation of the developing fragments. In an associative mechanism the importance of bond breaking will depend entirely upon circumstances and can range, in principle, from insignificance to dominance. The entering group is also present in the transition state and the influence of the one on the other must also be taken into account. Up till now it appears that for Pt(II) complexes the leaving group does not affect the nucleophilic discrimination factors significantly and only changes the intrinsic reactivity. Therefore a comparison of leaving group effects in a particular series of substrates will not depend upon the nature of the entering group. One of the most extensive series that has been studied has been [Pt dien X]$^+$ where the rate of replacement of X by pyridine decreases in the order $H_2O \gg Cl > Br > I > N_3 > -SCN > -NO_2 > -CN$. In this case, but not always, leaving group effects tend to parallel the strength of the metal–ligand bond. The leaving group effect can also depend upon the nature of the solvent. Although a considerable amount of work has been published on solvent effects no study has been sufficiently exhaustive or systematic to elucidate the problem fully.

5-6-4 Nature of the reaction centre

This is probably the most important aspect of any mechanistic discussion of four-coordinate planar substitution and yet at this time any discussion is inadequately covered by available data and much work remains to be done. A good deal of care must be exercised to ensure that the effects that are observed are truly due to the reaction centre. The sophistication of the approach can vary quite considerably from direct comparison of reaction rates to a detailed comparison of features of the intimate mechanisms.

(a) *Comparison of the reactivities of isovalent ions.* If we want to compare the reactivities of a series of complexes differing only in the nature of the central atom, such a comparison must obviously be restricted to isovalent ions. Since we are also restricted to d^8 ions, the best we can do is to compare triads, for example Ni(II), Pd(II), Pt(II), and Co(I), Rh(I), and Ir(I). The number of complete sequences that have been studied is almost vanishingly small, partly because it is no easy task to organize the same ligand environment and the same reaction and partly because the range of reactivity is so great. The general sequence is Ni $\gg$ Pd $\gg$ Pt in the ratio 10^7–10^8 : 10^5–10^6 : 1 and nobody has yet made adequate comparison for any other triad.

(b) *Comparison of non-isovalent ions.* Here it is impossible to have both identical ligand environments and identical charges. Quite often when the same ligand environment is subject to the same nucleophilic attack the course of the reaction will depend upon the nature of the central atom. Thus

$$[\text{Pt py Cl}_3]^- + \text{N}_3^- \longrightarrow [\text{Pt py Cl}_2\text{N}_3]^- + \text{Cl}^-$$
$$[\text{Au py Cl}_3]^\circ + \text{N}_3^- \longrightarrow [\text{Au Cl}_3\text{N}_3]^- + \text{py}$$

Even when the same reaction is observed, direct comparison of rate constants can be grossly misleading. For example, the complex $[\text{Au dien Cl}]^{2+}$ reacts with methanol some 30,000 times faster than $[\text{Pt dien Cl}]^+$ (Table 5-6), and indeed it is very common to find that an Au(III) complex is much more labile than the comparable Pt(II) one. However, the major part of the reactivity difference comes from the differences in the entropy

Table 5-6 Rates and activation parameters for the reaction $[\text{M dien Cl}]^{n+} + \text{Br}^- \rightarrow [\text{M dien Br}]^{n+} + \text{Cl}^-$

M		$k_{25°}$ in H_2O	$\Delta H^{\ddagger}$	$\Delta S^{\ddagger}$
Pt(II)	$n = 1$	$0\cdot 0053$ M^{-1}s^{-1}	14 kcal.mol^{-1}	-23 cal.deg^{-1}.mol^{-1}
Au(III)	$n = 2$	190 M^{-1}s^{-1}	13 kcal.mol^{-1}	-4 cal.deg^{-1}.mol^{-1}

of activation. This immediately indicates that arguments based entirely on the strength of the bonding are quite out of place and one must be greatly concerned with the nature of the interaction with the solvent.

(c) *Bond-making/bond-breaking relationships.* Although it is not yet possible to predict absolute values of reactivity it is possible to see patterns emerging when we consider the broader relationships between reactivity and the nature of the reaction centre. An important aspect is the relationship between bond making and bond breaking. Fig. 3-2 showed the two transition states that were required for an associative mechanism. Only when X = Y is it required that they have the same energy and at all other times the one that is higher will control the rate of reaction. The first transition state can be termed 'bond making' and the second 'bond breaking' but it must not be thought that the two processes are, of necessity, independent since a great deal will depend on the relative stability of the intermediate. A series of possibilities is shown diagrammatically in Fig. 5-7. (I) represents a truly synchronous I_a process in which there is no intermediate and where there is no possibility of separating the bond-making and bond-breaking aspects. In (II) the intermediate has only marginal stability and the two transition states are relatively close in energy, the separation of the bond-making and bond-breaking processes is only vestigial. In (III) the intermediate is considerably more stable although not yet likely to be seen, even from the kinetics. The first transition state is concerned mainly with the formation of the new bond and the leaving group functions mainly as one of the other four ligands in the complex. In the second transition state bonding with Y is virtually complete. In this type of process it is possible to consider bond-making relationships and bond-breaking relationships separately. In case (IV), the five-coordinate intermediate is of comparable stability to the reagents but less stable than the products. In this type of situation, the kinetics will

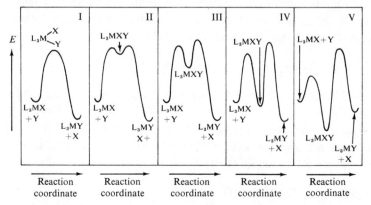

Fig. 5 – 7 Set of potential energy (E)—reaction coordinate curves for an associatively activated process showing increasing stabilization of the associative intermediate

certainly show the presence of the intermediate which may even build up in sufficient quantity to be directly observable. Case (V) represents the limit where five-coordination is more stable than four-coordination. Indeed here the tables have been turned and the planar four-coordinate species might be best looked upon as the intermediate in the dissociative reaction of a stable five-coordinate complex, that is

$$R_4MX \longrightarrow R_4M + X \quad \text{slow}$$
$$R_4M + X \longrightarrow R_4MX \quad \text{fast}$$
$$R_4M + Y \longrightarrow R_4MX \quad \text{fast}$$

If we make the assumption that the factors leading to stable five-coordination in other circumstances will also lower the energies of the unstable five-coordinate intermediates with respect to that of the four-coordinate planar form, we can say that the smaller the effective nuclear charge on the metal, the greater will be the relative stability of the five-coordinate intermediate.

Thus, Ir(I) > Pt(II) > Au(III)

and Ni(II) > Pd(II) > Pt(II)

Thus, the various cases in Fig. 5-7 do represent actual cases. (I) is typical of nucleophilic attack at a tetrahedral carbon where suitable orbitals for the formation of a fifth bond are not available. (II) represents the typical behaviour of Au(III) complexes where bond making and bond breaking are closely interconnected and nucleophilicity sequences depend upon the ligands present in the complex and upon the nature of the leaving group. (III) represents the typical behaviour of Pt(II) complexes where nucleophilicity can be considered independently of the nature of the leaving group. (II and III also represent the situation for nucleophilic substitution at an aromatic carbon centre where π-orbitals are available, or at tetrahedral Si, Ge, and P, where empty d orbitals can be used.) Very few examples of case (IV) have been characterized but the reaction

$$[LRh(SbR_3)Cl] + am \longrightarrow [LRhClam] + SbR_3$$

(L is a diolefin chelate such as cyclooctadiene, R is p-tolyl, and am is an amine) takes place at a rate that is dependent upon the nature, but not the concentration, of am. The rapid and reversible formation of a five-coordinate intermediate LRh(SbR$_3$)am.Cl has been invoked. It has also been suggested that the replacement of

$$(CH_3)_2C(OH)C \equiv C.C(OH) - (CH_3)_2 = ac$$

from [Pt Cl$_3$ac]$^-$ by 2,2'-bipyridyl follows the same mechanism.

5-7 The unimolecular mechanism

Having demonstrated that the normal mode of substitution at a square planar d^8 reaction centre is associative, it may be of some interest to see how the search for the dissociative mechanism is progressing. In general terms one would set about promoting a dissociative mechanism by one or more of the following three methods: (i) promote bond weakening; (ii) stabilize the intermediate of lower coordination number; (iii) prevent bond formation. In principle, it should be possible to achieve this either by electronic or steric effects.

(a) *Electron displacement.* Many cases where weakly electronegative ligands form a strong bond to the metal and cause extensive lengthening (and presumably weakening) of the *trans* metal-ligand bond have been documented. In some of these the kinetics of the substitution reactions have been studied and, while the reactivity is considerably increased, there is as yet no firm evidence for a change in kinetic form and possibly mechanism. A great deal of careful work is still required in this area before one can discount a change of mechanism. The stabilization of three-coordination in d^8 systems, other than by massive steric hindrance, has not yet been achieved.

(b) *Steric effects.* In general terms, the transition state for an associative reaction is more crowded than the ground state, whereas the transition state for a dissociative reaction is less crowded. If the ligands are chosen such that the ground state is somewhat congested it is to be expected that the associative reaction will be retarded, while the dissociative process may be accelerated since bond dissociation can relieve steric strain. Full success has been achieved by using *N*-substituted diethylenetriamine. The complexes of the parent ligand [M dien X]$^+$ (M = Pd, Pt) react by the usual A mechanism with the two-term rate law. In [M Et$_4$ dien X]$^+$, the ethyl groups block the top and bottom of the metal and make the associative mechanism difficult. This can be seen in the comparison of the reactions of [Pd dien Cl]$^+$ with some *N*-substituted analogues (Table 5-7). The reactivity is decreased by many powers of ten ($\sim 10^5$) and, in most cases, the k_2 term disappears. It has been suggested that the steric hindrance has prevented the bimolecular mechanism and that a reluctant and far less favourable dissociative mechanism has taken over by default. The changeover is not complete and it should be realized that stable five-coordinate complexes of Co(II) and Ni(II) with Et$_4$ dien are well known. Indeed this ligand is used to stabilize five-coordination where the normal behaviour would lead to six-coordinate octahedral complexes. The nucleophiles that remain effective are interesting. Only thiosulphate, and in some cases, hydroxide and thiourea, are able to rouse sufficient

Table 5-7 Rate constants for the substitution reactions of $[Pd(AAA)X]^+$ in water at 25°

AAA[a]	X	$k_1(s^{-1})$	Y	$k_2(M^{-1}s^{-1})$
dien	Cl	100		
Me$_5$dien	Cl	0·240	$S_2O_3^=$	12·6
Me$_5$dien	Cl		$SC(NH_2)_2$	1·4
			OH^- and others	0
Et$_4$dien	Cl	0·002[b]	OH^-	0·06
			others	0
Et$_4$dien	Br	0·0015	$S_2O_3^=$	5·9
			OH^-	0·04
			others	0
MeEt$_4$dien	Cl	0·00065	$S_2O_3^=$	0·052
			OH^- and others	0

[a] dien = $H_2NCH_2CH_2NHCH_2CH_2NH_2$
Me$_5$dien = $(CH_3)_2NCH_2CH_2N(CH_3)CH_2CH_2N(CH_3)_2$
Et$_4$dien = $(C_2H_5)_2NCH_2CH_2NHCH_2CH_2N(C_2H_5)_2$
MeEt$_4$dien = $(C_2H_5)NCH_2CH_2N(CH_3)CH_2CH_2N(C_2H_5)_2$
[b] The rate constant for the Pt(II) analogue is $8·5 \times 10^{-6} s^{-1}$ at 80°C

enthusiasm from the complex to generate a rate dependence. The sulphur donors are probably effective because they are strong nucleophiles and the hydroxide is so because it deprotonates the secondary nitrogen and produces a dissociatively labile conjugate base. This behaviour vanishes when no acidic protons are available and is very reminiscent of the base catalysed hydrolysis in octahedral substitution.

Another potential source of considerable hindrance can be found in complexes of the type *cis*- and *trans*-[Pt(PEt$_3$)$_2$RCl], where R is an *o*-substituted phenyl group. The original work indicated that the effect of placing methyl groups *ortho* to the Pt—C bond was much more serious when R was *cis* to the leaving group (Table 5-8). This is fully in accord with the anticipated form of the trigonal bipyramid and there the ortho substituents would cause most interference (Fig. 5-8). In the *trans* complex there did not seem to be any interference with the regular kinetic form and so the associative mechanism is retained. More detailed studies in recent years have shown that the situation is more complicated than was originally suggested and that many nucleophiles are inactive in this system. Whether the absence of a k_2 term is indicative of a dissociative mechanism, as in the case of the *N*-substituted dien complexes, is still being questioned. The isomerization of *cis*-[Pt(PEt$_3$)$_2$(o-tolyl)Cl] [see Section 5-8(iv)] indicates that the dissociative mechanism, while obviously present, only accounts for a small part of the substitution that occurs by way of the nucleophile independent path. A similar type of effect has been

The unimolecular mechanism

Table 5-8 Rate constants for the reaction $[Pt(PEt_3)_2RCl] + py \rightleftharpoons [Pt(PEt_3)_2Rpy]^+ + Cl^-$ in methanol

R		$10^4 k_1 (s^{-1})$		$10^4 k_2 (M^{-1}s^{-1})$	
cis, C$_6$H$_5$		380	(0°)	—	
cis, 4-CH$_3$-C$_6$H$_4$		500	(0°)	—	
cis, 2-CH$_3$-C$_6$H$_4$		0·87	(0°)	—	
cis, 2,4,6-(CH$_3$)$_3$-C$_6$H$_2$		0·0042	(0°)	—	
trans, C$_6$H$_5$		0·33	(25°)	75	(25°)
trans, 2-CH$_3$-C$_6$H$_4$		0·067	(25°)	17	(25°)
trans, 2,4,6-(CH$_3$)$_3$-C$_6$H$_2$		0·017	(25°)	3·7	(25°)

Note added in proof: Recently completed work by Faraone et al. indicates that pyridine does not attack these complexes directly. The k_2 terms should be ignored.

found with *trans*-[PtCl$_2$(PEt$_3$)(NHEt$_2$)] where, as was mentioned in Section 5-4(iii), the k_2 term was not observed in methanol. Any thought that this might indicate a dissociative mechanism is dispelled when the solvent is changed to the non-coordinating hexane, where the k_2 term appears and the k_1 term vanishes.

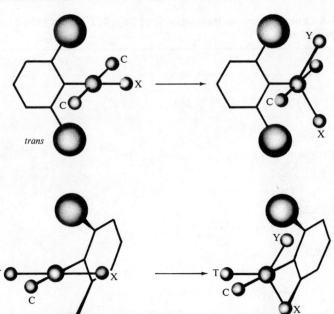

Fig. 5-8

5-8 Stereochemistry of substitution in planar four-coordinate complexes

In general, substitution in these systems takes place with complete retention of configuration and it has been shown how this would be the expected consequence of the trigonal bipyramidal intermediate. There are, nevertheless, numerous cases of substitution that lead to steric change and many other cases of isomerization. Only a few have been examined in sufficient detail to warrant any mechanistic discussion. Four major possibilities are worth discussing:

(i) *A multistage reaction, each step of which involves retention of configuration*. This has been demonstrated for the *cis* ⇌ *trans* isomerization of $[Pd\ am_2Cl_2]$, which is catalysed by excess amine 'am' at rates that reflect its nucleophilicity. The suggested mechanism is

$$\underset{am}{\overset{Cl}{am-Pd-Cl}} + am \longrightarrow \underset{am}{\overset{am}{am-Pd-Cl}} \overset{+Cl^-}{\underset{+Cl^-}{\rightleftharpoons}} \underset{am}{\overset{am}{Cl-Pd-Cl}} + am$$

Sec. 5-8   Stereochemistry of substitution   67

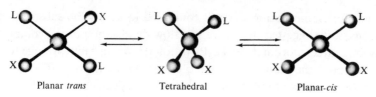

Fig. 5 – 9 Intramolecular *trans* ⇌ *cis* isomerization in a four-coordinate planar complex by way of a tetrahedral transition state

(ii) *Intramolecular isomerization in the four coordinate state.* Many four-coordinate Ni(II) complexes of the type NiL_2X_2 exist in solution as an equilibrium mixture of diamagnetic planar and paramagnetic tetrahedral species. The composition can be measured by ^1H n.m.r. and in favourable cases the rate of interconversion, which is normally quite rapid, can be measured. Since, in the tetrahedral form, both ligands L and both ligands X are symmetrically equivalent, this provides a suitable path for *cis-trans* changes in the planar form (Fig. 5-9). The tetrahedral high spin form is energetically far less accessible for the platinum(II) analogues and no intramolecular isomerizations of this sort have been found there yet. However, the photochemically induced *cis* ⇌ *trans* isomerization of [Pt(glycine)$_2$] is completely intramolecular and a tetrahedral excited state has been suggested.

(iii) *Intramolecular isomerization in the five-coordinate intermediate.* Stable five-coordinate species, such as $Fe(CO)_5$ and PF_5 are known to undergo rapid 'pseudorotation' in which axial and equatorial positions are exchanged. If the lifetime of the five-coordinate intermediate in substitution could be made long enough for some sort of pseudorotation to take place, isomerization could be the result. It has long been known that the *cis* ⇌ *trans* isomerization of [Pt(Et$_3$P)$_2$Cl$_2$] (and similar species) is catalysed by free phosphine and since reaction is most facile in non-ionizing solvents it has been suggested that release of chloride, as required by mechanism (i) above, does not take place. Pseudorotation in the [Pt(PEt$_3$)$_3$Cl$_2$] intermediate is invoked. This must be rather specific because recent work has suggested that the catalyst phosphine *does not* exchange with the coordinated phosphine. Isomerizations and substitutions with steric change are indeed more common in the reactions of the more 'covalent' d^8 complexes, where five-coordination is likely to be most favoured.

(iv) *Dissociative substitution with a three-coordinate intermediate.* It has recently been shown by Faraone *et al.* that *cis*-[Pt(PEt$_3$)(*o*-tolyl)Cl] isomerizes to the *trans* isomer. This reaction is not catalysed and in fact is suppressed to some extent by added chloride. Since the first-order rate constant for isomerization is small compared to k_1 for the substitution reactions, it is suggested that there is a slow dissociation of the complex to give the three-coordinate $Pt(Et_3P)_2(o\text{-tolyl})^+$ which can rearrange to a form that gives the *trans* product when chloride is reattached. It is hoped that more examples of this behaviour will be discovered.

Problems

5-1 Making use of the *trans*-effect sequence, show how you would attempt to synthesize the three isomers of [Pt py NH_3NO_2Cl] starting from K_2PtCl_4.

5-2 Outline and discuss the circumstances under which stereochemical change is found in four-coordinate planar platinum(II) complexes.

5-3 Write an account of the *trans*-effect in four-coordinate planar d^8 complexes.

5-4 Discuss the evidence that proves that the k_1 term found in the rate law for square planar substitution relates to a bimolecular solvolysis. Why is the test of Olcott and Gray inapplicable to the examination of cryptosolvolysis in octahedral cobalt(III) complexes? (Do not attempt the second part before reading Chapter 7.)

5-5 Write an account of the factors that are important in determining the nucleophilicity of a reagent towards platinum(II) complexes.

5-6 Discuss the occurrence of dissociative mechanisms in the substitution reactions of four-coordinate planar complexes and comment critically upon the evidence presented.

Bibliography

Basolo, F. and R. G. Pearson, The *trans*-effect in metal complexes, *Prog. Inorg. Chem.* (Ed. F. A. Cotton), 1963, **4**, 381.

Cattalini, L., The intimate mechanism of replacement in d^8 square-planar complexes, *Prog. Inorg. Chem.* (Ed. J. O. Edwards), 1970, **13**, 263.

6 Substitution in five-coordinate systems

6-1 Introduction

Not much more than a decade ago, stable five-coordination was considered to be a scientific rarity when found outside a narrow, restricted area of the periodic table. Since then a great deal of work has been published and nowadays one can become very blasé about the discovery of yet another five-coordinate system. In contrast to this, five-coordination has always been of considerable importance in simple substitution reactions since most of the intermediates and transition states were five-coordinate. Thus the associative mechanism which is usual for substitution in tetrahedral and planar four-coordinate substrates and, as will be seen, the dissociative mechanism which is the norm for octahedral substitution will all give rise to five-coordinate transition states or intermediates. A major area of mechanistic study in recent years has been concerned with the phenomena of stereochemical non-rigidity or pseudorotation. The relative ease with which many five-coordinate compounds can undergo intramolecular topological changes has attracted many investigators and much work has been done. In contrast to all of these mechanistic features, actual kinetic studies of substitution reactions of five-coordinate compounds are very rare indeed at the moment.

6-2 Occurrence of five-coordination

Examples of stable five-coordinate compounds can be found in most parts of the periodic table although the areas where it is the norm, rather than the exception, are few.

6-2-1 P-block compounds

Five-coordination can occur with the following valence shell electron configurations:

(a) Five bonding pairs of electrons will lead to compounds such as PF_5 and the other group V halides which are five-coordinate in the gas phase, $P(CH_3)_2F_3$, and so on. Even here, the tendency is to change to higher, and sometimes lower, coordination numbers by sharing or transferring a ligand (for example, PCl_5 (gas) $\rightarrow PCl_4^+ + PCl_6^-$) in the solid state and in ionizing solvents, $(SbF_5)_n$ is polymeric with fluorine bridging, and so on. Isoelectronic compounds such as SiF_5^- are rare and generally stabilized only in certain ionic lattices. Complexes of group III elements occasionally adopt five-coordination, for example, $GaCl_3py_2$ and $(Et_3N)_2AlH_3$, but these are uncommon when compared to the analogous four- or six-coordinate species.

(b) Six bonding pairs of electrons, including one double bond are found in compounds such as SOF_4 and $TeOF_4$, and this section can be expanded to seven bonding pairs of electrons with one triple or two double bonds, as in ClO_2F_3 and IO_2F_3.

(c) A 12-electron valence shell with five bonding pairs and one lone pair will produce a square pyramidal compound such as IF_5, TeF_5^-.

6-2-2 D-block compounds

Three main areas of five-coordination exist in transition metal chemistry.

(a) The higher oxidation states of the earlier transition elements bear a superficial resemblance to the corresponding P-group element in the other sub-group and similar relationships between electron configuration and five-coordination exist, for example, VF_5, $VO(acac)_2$. In high oxidation states, the presence of one or two non-bonding d electrons does not appear to interfere with the stereochemistry. The ability of these species to act as Lewis acids and increase their coordination to six ranges from reasonably strong to exceptionally strong.

(b) Five-coordination is common when the central atom has the d^8 configuration since a 'noble gas' electronic structure would require five bonding pairs of electrons to go with the eight non-bonding ones. The interplay of the various features controlling coordination number, coordination geometry, and spin multiplicity in d^8 systems has already been discussed in Chapter 5.

(c) The use of multidentate ligands has greatly increased the number of known examples of stable five-coordination, especially of compounds of the elements of the first row of the transition series. Quite often the

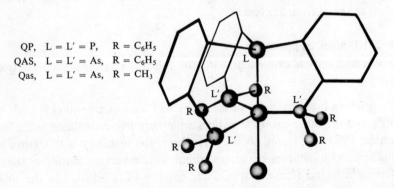

QP, L = L' = P, R = C_6H_5
QAS, L = L' = As, R = C_6H_5
Qas, L = L' = As, R = CH_3

Fig. 6–1 Trigonal bipyramidal complex of a 'tripod' quadridentate ligand showing how the trigonal symmetry of the ligand forces the configuration in the complex

cause is steric, either by preventing the usual six-coordination, as in [Co(Et$_4$dien)Br]X, or by preventing the usual geometry associated with a more common lower coordination, as in the case of the complexes of the 'tripod' quadridentate ligands, such as tris(*o*-diphenylphosphinophenyl)-phosphine = QP, which forms trigonal bipyramidal five-coordinate complexes, such as, [Pt(QP)Br]ClO$_4$ (Fig. 6-1).

6-3 General remarks on mechanism

Any attempt to predict the relative expectation of associative or dissociative mechanisms must take into account the amount of coordination unsaturation and the accessibility of higher and lower coordination number states. For example, it is well known that phosphorus pentalides and their derivatives have a marked tendency to act as Lewis acids and increase their coordination number to six. This tendency, for the pentahalides PX$_5$, decreases markedly in the sequence F > Cl > Br. At the same time, tetrahedral species such as PX$_4^+$ are also known, their stability increasing along the sequence F < Cl < Br. Therefore a duality of mechanism is possible since both modes of activation are accessible. Much will depend upon the nucleophile and the nature of the groups to the reaction centre and the reactions will probably be very fast.

On changing from phosphorus to arsenic and antimony the tendency to increase the coordination number becomes dominant and associative paths ought to be predominant. Thus, whereas a solution of PCl$_5$ in acetonitrile contains mainly the tetrahedral PCl$_4^+$ and the octahedral PCl$_6^-$ ions, antimony pentachloride gives octahedral *trans*-[SbCl$_4$(CH$_3$CN)$_2$]$^+$ and SbCl$_6^-$ ions only.

The low-spin, five-coordinate d^8 complexes represent a coordinatively saturated system if the 'nine-orbital' rule is controlling the behaviour and it is reasonable to believe that dissociative activation will be the order of the day. The facility of this mechanism will increase as the four-coordinate planar arrangement gains stability at the expense of the five-coordinate system. Associative mechanisms need not be ruled out, especially in high-spin systems or when the bonding is somewhat more electrovalent.

Unfortunately, while it is possible and even pleasurable to speculate extensively, the amount of hard fact that can be used to test these speculations is very small indeed. Studies of simple substitution, uncomplicated by oxidative addition or pseudorotation are few. The five-coordinate complexes of Ni(II), Pd(II), and Pt(II) ought to present a very fertile ground since, as was pointed out in Section 5-6-4(c), there ought to be a reciprocal relationship between five-coordinate intermediates and transition states for associative substitution in four-coordinate planar and tetrahedral substrates and four-coordinate planar and tetrahedral transi-

72 Substitution in five-coordinate systems Ch. 6

tion states and intermediates for dissociative substitution reactions of five-coordinate complexes (Fig. 6-2). In certain cases, the sterically facile associative mechanism need not be ruled out. The examples shown in the figure are but a selection of the many possibilities. The reader should try to see how many other paths he can draw.

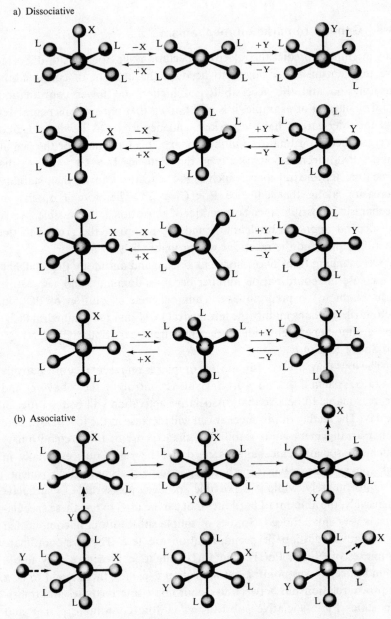

Fig. 6 – 2 Selection of possible pathways for substitution in five-coordinate complexes

6-4 Substitution in five-coordinate Ni(II), Pd(II), and Pt(II) complexes

In general, these reactions are very fast and much more has been done to study the four-coordinate ⇌ five-coordinate equilibria than to examine the kinetics and mechanism. However, substitution reactions of five-

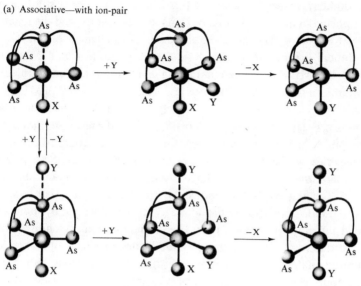

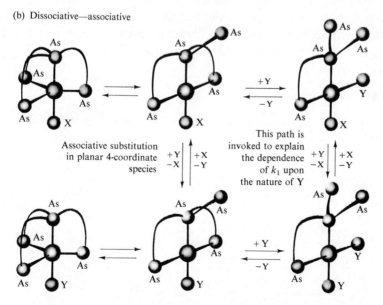

Fig. 6–3 Alternative paths for the substitution reactions of [M L X]⁺

coordinate complexes, where the requirements of a multidentate ligand are rigorous, can be of considerable interest.

The kinetics of the reaction

$$[MLX]^+ + Y \longrightarrow [MLY]^+ + X$$

where M = Ni, Pd, Pt; L = a trigonally symmetrical ligand such as tris(*o*-diphenylarsinophenyl)arsine (QAS in Fig. 6-1) or tris(*o*-dimethylarsinophenyl)arsine (Qas in Fig. 6-1), have been studied and the reactions are relatively slow. The steric requirements of the ligand force a trigonal bipyramidal arrangement with X in an axial position and completely prevent the ML^{2+} fragment assuming a planar (or even tetrahedral) configuration (Fig. 6-2a). The rate law, rate = $(k_1 + k_2[Y])[MLX^+]$, is reminiscent of the substitution reactions of four-coordinate planar complexes, except that k_1 is very dependent upon the nature of the entering nucleophile. The dependence upon the nature and concentration of Y indicates clearly that a simple dissociative activation is out of the question. Two alternative mechanisms have been proposed and are shown in Fig. 6-3. The first involves a bimolecular attack leading to a six-coordinate transition state or intermediate and the k_1 term is explained by some weird parallel path involving an ion-pair. The second mechanism requires a rapid five-coordinate $\rightleftarrows$ four-coordinate equilibrium in which one arsenic of the multidentate becomes detached, with rapid bimolecular substitution in the four-coordinate species. The k_1 term, in part, arises from the possibility of displacing a second arsenic. The author of this book is partisan with respect to this problem and cannot be relied upon to give a fair assessment of the relative merits of the two mechanisms but it is clear that the nucleophile dependent k_1 term presents problems and the last word on the mechanism has yet to be uttered.

6-5 Conclusion

Perhaps in five years time speculation will be replaced by factual information.

Problems

6-1 Write an essay on the importance of five-coordination in the field of inorganic reaction mechanisms.

6-2 Why is the apical bromide in bromodi(o-phenylenebisdimethylarsine)-platinum(II) cation so much more labile than the axial bromide in bromo [tris(o-dimethylarsinophenyl)arsine]platinum(II) cation?

Bibliography

Muetterties, E. L. and R. A. Schunn, Penta-coordination, *Quart. Revs.*, 1966, **20**, 245

Nyholm, R. S. and M. L. Tobe, *The Stabilization, Stereochemistry and Reactivity of Five-coordinate Complexes* (Essays in Coordination Chemistry), pp. 112-127. Birkhäusen-Verlag, Basel, 1964.

7 Substitution at six-coordinate reaction centres

7-1 Introduction

Of the possible geometries for six-coordination only the octahedral or distorted octahedral configuration has received any study. If one discounts the very large number of variations on a tetrahedral theme that constitutes much of organic chemistry, it is reasonably fair to claim that the octahedron is the commonest, and certainly the most ubiquitous, geometry of chemistry. It represents a very stable arrangement for purely electrostatic interaction between a coordination acceptor centre and a ligand whose radius ratio sits within the right range; it is a stable arrangement, either regular or irregular, for many partly electrovalent systems; it is a standard arrangement for the heavier non-transition elements and is the strongly favoured arrangement for the low-spin d^6 configuration of the transition elements. The important areas are summarized in Table 7-1 which is by no means comprehensive. Covering as it does such a very wide range of oxidation states and electron configurations it might have been

Table 7-1 Occurrence of the octahedral geometry

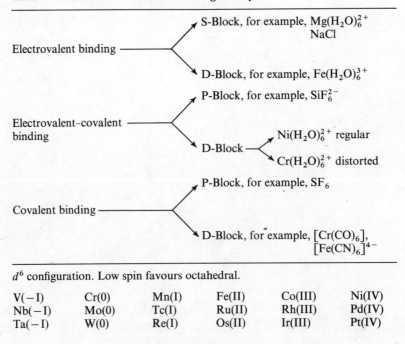

d^6 configuration. Low spin favours octahedral.

V(−I)	Cr(0)	Mn(I)	Fe(II)	Co(III)	Ni(IV)
Nb(−I)	Mo(0)	Tc(I)	Ru(II)	Rh(III)	Pd(IV)
Ta(−I)	W(0)	Re(I)	Os(II)	Ir(III)	Pt(IV)

expected to show a multitude of different mechanisms but, as will emerge in the course of this chapter, the mechanistic pattern is extremely consistent although not necessarily straightforward and although we are dealing with such a wide range of source material it is possible to speak of a characteristic octahedral mechanism and to search, with singular lack of success, for marked departures from the typical behaviour.

7-2 Quasi-theoretical arguments

7-2-1 Valence bond considerations

Much has been said in the past about the application of valence bond arguments to the problem of mechanism. Taube's famous review in 1953 is now part of history and having served the cause well is in honourable retirement. Two major points were made at that time which are still of great importance. The first drew attention to the difference between kinetic inertness and thermodynamic stability and the need to specify clearly which of these was being discussed in any particular case. Sad to say many people who should have known better still perpetuate this error. The second pointed out the relationship between lability and the accessibility of a particular reaction path. Both these points have been discussed in Chapter 1. Applied to octahedral complexes the Taube hypotheses used the fashionable but false assumption that d^2sp^3 or inner d-orbital complexes were covalent and sp^3d^2 or outer d-orbital complexes were ionic. With the inner orbital complexes, d^0, d^1, and d^2, configurations left an empty inner d-orbital which could provide a facile path for an associative process and hence lead to the labile complexes. A more up-to-date view here would admit that the octahedral d^0, d^1, d^2 (and even d^3 and d^4 if spin pairing can occur) complexes are coordinately unsaturated and in the right circumstances stable coordination numbers greater than six are possible. Eight-coordinate $TiCl_4(diars)_2$ and seven-coordinate $[Mo(diars)_2(CO)_2X]^+$ are just two of many examples in this area of the periodic table. An associative mechanism, therefore, might be expected in this area. The outer orbital complexes on the other hand, wrongly thought to be ionic, were consequently expected to react dissociatively and the greater the effective charge on the central atom the smaller the lability. Taube pointed out that there was a very marked decrease in lability along the sequence $AlF_6^{3-} > SiF_6^{2-} > PF_6^- > SF_6$. The conclusions for the transition elements are summarized in Table 7-2.

7-2-2 Crystal field theory

Basolo and Pearson in the first edition of their definitive text on inorganic reactions in solution in 1958 were attracted by the then fashionable crystal field theory and attempted a semi-quantitative rationalization of reactivity on the basis of the difference between the crystal field stabilization energy of the reagent and the transition state, assuming that this contributed directly to the activation energy of the process. The reader is recommended to go to the original for further information (F. Basolo and R. G. Pearson, *Mechanisms of Inorganic Reactions*, 2nd Edition, John Wiley, New York, 1967, pp 145-158) but bearing in mind the facts that (i) a purely electrostatic model is assumed; (ii) the crystal field stabilization energy is just one of a number of energetic terms that must be considered; (iii) the calculations relate to regular geometries and take no account of any change in effective crystal field strength of the remaining ligands on going from the

Table 7-2 Valence bond model for octahedral complexes. Electron configurations leading to inert complexes are shaded

Inner orbital complexes $(n-1)d^2\, ns\, np^3$	Outer orbital complexes $ns\, np^3\, nd^2$

ground state to the transition state, and so on, it is surprising that any correlation between the fact and theory exists, but exist it does. The approach is most successful when it is applied to the relative reactivities or, better still, activation energies for a particular reaction of a series of complexes that differ only in the nature of the central atom. Thus it is possible to order the lability of the aquotransition metal ions in terms of their electron configurations, to explain why iridium(III) complexes are much more inert than cobalt(III) complexes, to postulate that it is always easier in respect of loss of crystal field stabilization energy to generate a square pyramidal five-coordinate intermediate rather than a trigonal bipyramidal one. At this point the arguments become dangerous and one might be reminded that, according to crystal field theory, tetrahedral nickel(II) species ought not to exist, but they do exist and they can be thermodynamically stable. It was, in fact, the dogmatic statement that 'crystal field theory has proved that a tetrahedral d^8 configuration cannot exist' that led to a flurry of research which soon produced many unequivocal examples of regular tetrahedral nickel(II) complexes. Rather than discuss the matter any further in isolation it will be better to consider the examples in their appropriate context.

7-2-3 Extrapolation of information about stable five- and seven-coordination

As has already been mentioned above, the occurrence of stable seven- and higher coordination numbers is restricted mainly to transition element complexes with d^0, d^1, d^2, d^3, and d^4 electron configuration and becomes increasingly likely as the bonding becomes more covalent. Octahedral complexes with these configurations might be looked upon as coordinatively unsaturated. In an essentially electrovalent situation, coordination numbers greater than six can be obtained, as in $[\text{Fe EDTA H}_2\text{O}]^-$ which is a seven-coordinate complex of the d^5 Fe(III). With larger ions, such as the lanthanides, coordination numbers up to nine are well known. Coordination unsaturation is also found in the heavier P-block elements as for example $\text{IF}_6^+ \rightarrow \text{IF}_7$. Simple substitution reactions are not common in this area and oxidation-reduction processes, $\text{IF}_7 \rightleftharpoons \text{IF}_5 + \text{F}_2$ would dominate.

Five-coordinate complexes can be stabilized from a normally octahedral situation under suitable conditions. Thus, steric hindrance in high-spin Ni(II), Co(II), Fe(II), Cu(II), and Zn(II) (the last two have no choice about spin multiplicity) using the right sort of ligands will lead to stable five-coordinate complexes. A delocalized electronic arrangement in the ligand system, as in the cobalamines, cobaloximes, and certain quadridentate Schiff base complexes of cobalt(III) can lead to a planar four-coordinate $\rightleftharpoons$ tetragonal pyramidal-five-coordinate $\rightleftharpoons$ octahedral-six-coordinate equilibrium situation. In these cases it appears that strong σ-bonded ligands can weaken the ligand *trans* to themselves sufficiently to stabilize five-coordination.

7-2-4 General conclusions
All that can be said is that, unlike the d^8 planar four-coordinate substitution, no one mechanistic path stands out above all others. On the other hand one can predict (with safety from behind the barricade of a large amount of factual data) that the dissociative type of reaction will be more probable.

7-3 General remarks on the distribution of experimental evidence

There are two main approaches that can be adopted to the study of substitution reactions of octahedral complexes:

(i) *Studies of labile complexes.* The general methodology of study of such complexes involves taking a system at equilibrium, applying a perturbation, and then examining the relaxation. This can be done in a variety of ways from stopped flow, temperature jump, and other single relaxation studies to n.m.r., ultrasonic absorption, and other periodic perturbations. This approach enables a very wide range of reaction centres to be studied in order to provide a restricted amount of information about each. The major drawback is that the investigator is at the mercy of the equilibrium requirements of the system and cannot control the nature of the substrate adequately.

(ii) *Studies of relatively inert complexes.* These are somewhat restricted in their availability and so the breadth of the alternative approach is lacking. On the other hand, the scope for modifying the immediate environment of the reaction centre is considerable and the effect of varying the nature and position of the other ligands in the complex upon the kinetics of the reaction allows a study in depth to compensate the lack of breadth. In addition, substrates can be signposted and detailed stereochemical evidence obtained. This approach closely resembles that which has already been discussed for four-coordinate planar d^8 complexes and, indeed, is the classical approach applied to the study of organic reaction mechanisms.

7-4 Kinetics and mechanism of the reactions of solvated metal cations

Apart from $Cr(H_2O)_6^{3+}$ and $Rh(H_2O)_6^{3+}$, the aquo metal cations exchange their ligands with solvent water with great rapidity. In the past this had led to difficulty in distinguishing between coordination and solvation but

Table 7-3 Rate constants and enthalpies of activation for the exchange of a single water molecule in hexaaquo high-spin transition metal cations (25°, s^{-1}, activation enthalpies, kcal.mol^{-1}, in parentheses)

d^0	d^1	d^2	d^3	d^4	d^5	d^6	d^7	d^8	d^9	d^{10}
Sc^{3+} [2×10^7]$^{(a)}$			V^{2+} [1·2×10^2] [(15·3)]	Cr^{2+} 8·3×10^9	Mn^{2+} 3·1×10^6 (8·1)	Fe^{2+} 3×10^6 (8)	Co^{2+} 2·5×10^6 (11·5)	Ni^{2+} 3·6×10^4 (12·3)	Cu^{2+} 7·4×10^9	Zn^{2+} [2×10^7]
		V$^{3+\,(b)}$ 3×10^3	Cr^{3+} 5×10^{-7} (26·7)		Fe^{3+} [2·6×10^3]	Co^{3+} [~10^3]				Ga$^{3+\,(c)}$ 2·0×10^3 (6·3)
						Rh$^{3+\,(d)}$ 4×10^{-8} (33)				In^{3+} [2×10^5]

(a) Values in square brackets are estimated from the rates of complex formation.
(b) This value is based on the assumption that the mechanism of complex formation is the same as for Fe^{3+}.
(c) Note that, for Al^{3+}, k = 1·33 × 10^{-1} s^{-1} (27 kcal.mol^{-1}).
(d) Low spin.

nowadays, due mainly to the application of nuclear magnetic resonance techniques, it is possible not only to measure the number of solvent molecules coordinated to a metal ion, but also to determine the average time they spend in the coordination shell. Some rate constants for water exchange are collected in Table 7-3. The most obvious point is the very marked dependence of rate on electron configuration. There is a parallel with the loss of crystal field stabilization energy, the rates reaching a minimum at d^3 and d^8 for high-spin complexes, but departures in $d^4(Cr^{2+})$ and $d^9(Cu^{2+})$ are most probably due to the strongly tetragonally distorted nature of these complexes and represent the great lability of the weakest bound water molecules. In general, for a given electron configuration, the higher the oxidation state the less labile the complex. This is what would be expected from any consideration of bond strength. The rate also increases markedly with increasing ionic radius (c.f. Al^{3+} and Ga^{3+}).

A great deal of work has been done on the kinetics of the complex formation reactions of these metal ions. In general, the reactions are studied under reversible conditions and the rate constants for the forward (complex formation) and reverse (solvolytic) processes can be obtained. The techniques of microwave absorption, ultrasonic absorption, n.m.r. line broadening, temperature jump, stopped flow, and, in rare cases, classical methods for slow reactions, have all been applied to these problems. In spite of the wide range of reactions and reaction rates, a common mechanistic pattern emerges (it even extends to labile complexes where the coordination number is not six, as in tetrahedral $Be(H_2O)_4^{2+}$ and nine-coordinate $M(H_2O)_9^{3+}$ lanthanide ions). In its simplest form complex formation is characterized by two relaxation processes. Step 1 is a diffusion controlled process, τ (half-life for relaxation) $\sim 10^{-8} \to 10^{-10}$ s, and this is dependent mainly upon the charges of the species and the nature of the solvent (the behaviour is not restricted to aqueous solution). Step 2 is slower, almost independent of the nature and the concentration of the ligand, but very sensitive to the nature of the metal ion and the solvent. These two relaxations are interpreted as follows, as in the case of the formation of $FeCl(H_2O)_5^{2+}$.

Step 1. $Fe(H_2O)_6^{3+} + Cl^- \rightleftharpoons Fe(H_2O)_6^{3+} \ldots Cl^-$

Rapid, diffusion-controlled formation of an outer sphere complex, in this case it is an ion-pair.

Step 2. $Fe(H_2O)_6^{3+} \ldots Cl^- \rightleftharpoons Fe(H_2O)_5Cl^{2+} \ldots H_2O$

A slow interchange process in which the chloride ion has to wait for a water molecule to vacate a coordination site before it stands a chance to slip in.

In general, the actual rate of the interchange process is similar to the rate of solvent exchange in the absence of the ligand. Occasionally there are departures from this simple scheme but generally they are found in unusual circumstances. Thus, in the reactions of Cu^{2+}, where water exchange is extremely fast ($\sim = 10^{-8}$ s), the rate of formation of stable chelate complexes can be controlled by other features, such as ring closure. The reaction

$$Cu^{2+}aq + NH_2-\underset{CH_3}{CH}.CH_2COO^- \rightleftharpoons Cu^+\begin{matrix} NH_2CH.CH_3 \\ \diagdown \\ CH_2 \\ \diagup \\ O-C \\ \| \\ O \end{matrix}$$

is about 100 times slower than expected because of the relative slowness of ring closure.

The complex formation reaction is thus, typically, an I_d mechanism and, with certain modifications that are a matter of degree rather than principle, this is the typical octahedral substitution mechanism.

7-5 General kinetic features of octahedral substitution

At this stage it is convenient to summarize what has been said about inert systems in order to give a preview of what will be called the typical octahedral kinetic behaviour. The following major points are of importance.

7-5-1 Nature of the entering group

The rate of substitution is rarely very strongly dependent on the nature of the entering group and when effects due to pre-equilibrium ion association, proton transfer, and other phenomena are taken into account, it is not greatly dependent upon its concentration either. This would imply that a 'd' intimate mechanism is considerably more plausible than an 'a' mechanism. From time to time people assign bimolecular mechanisms and it is not unlikely that many of the reactions sit close to the 'a'-'d' borderline, but it is fair to say that the marked sensitivity of rate on the nature of the incoming group that is normal for four-coordinate planar d^8 substitution and which is expected if there is significant bonding with the entering group in the rate-determining transition state is, up till now, totally absent in simple octahedral substitution.

7-5-2 Characterization of a reactive five-coordinate intermediate

Although the kinetics indicate dissociative activation the second requirement for a D mechanism is often lacking. The identification of a labile

Table 7-4 Some octahedral complexes that have been shown to undergo reaction by way of a D mechanism (X is the leaving group)

$[Co(CN)_5X]^{3- \text{ or } 2-}$	trans-$[Co(dmg)_2RX]^-$ (a)
$[Co(CN)_4SO_3X]^{4-}$	when R is a strong trans-effect ligand
trans-$[Co(NH_3)_4SO_3X]$	$[RhCl_5H_2O]^{2-}$ (b)
trans-$[Co\ en_2SO_3X]$	Any Co(III) complex containing an amido
trans-$[Co(cyclam)ClH_2O]^{2+}$ (b)	(—ṄHR) group, for example, $[Co(NH_3)_4NH_2X]^+$

(a) This is the prototype for a wide range of complexes with conjugated ligand systems which may be macrocyclic (for example, cobalamins) and which form stable metal-ligand bonds with strong trans-effect ligands such as CH_3—, H—, —PR_3 and so on. dmg = dimethylglyoximate.
(b) Leaving group is H_2O. cyclam = 1,4,8,11-tetraazacyclotetradecane.

five-coordinate intermediate generally is associated with its behaviour on consumption. This can be summarized under the headings (a) competition, (b) discrimination, and (c) stereochemistry, and should be quite independent of its mode of formation. A selection of substrates that seem to obey the rules and generate identifiable five-coordinate intermediates are given in Table 7-4 and their reactions are assigned a D mechanism but, on the whole, such behaviour is unusual.

7-5-3 Role of the solvent

In a process where the substrate is unable to discriminate between possible reagents mob rule wins and the one in maximum supply dominates the proceedings. The solvent is generally present in large excess over the other reagents and when it is a potential ligand it can dominate the substitution processes. Even as a poor ligand it tends to win, but the weakness of its bond to the metal causes it to dissociate repeatedly thus giving the other species a chance. In all cases competition between the solvent and the other potential ligands for a place in the inner solvation shell is of considerable kinetic importance since this is a means whereby the reagent can boost its effective concentration with respect to that of the solvent.

7-5-4 Associative mechanism

In spite of many false alarms, the A mechanism has not yet been adequately demonstrated in octahedral systems. As has already been said, in most cases where a strong rate dependence on the incoming group exists, the effect can be blamed on pre-equilibrium interaction between reagents. In other cases competing processes, as in the competition with solvent or the displacement of multidentate ligands, can take the blame and sometimes it turns out that the act of substitution has not been taking place at the octahedral centre at all. A case in point is that of $[Si(acac)_3]^+$ which, if resolved, can lose its optical activity at rates that are first-order with respect to the concentration of added nucleophiles and whose second-order rate constants cover a range of 10^9 from the weakest (H_2O) to the strongest (OOH^-). This oft-quoted example of the A mechanism in octahedral substitution is somewhat undermined by the fact that in the similar $[Si(bzac)_3]^+$ (bzac = benzoylacetonate) it is the C—O and not the Si—O bond that is broken.

One promising area is that of the coordinatively unsaturated octahedral complexes and there is some indication that complex formation reactions of $[V(H_2O)_6]^{3+}$ (d^2) have some unusual characteristics in the enthalpies and entropies of activation which might be interpreted in terms of an associative mechanism. Recent measurements have indicated that the water exchange reactions of $[Cr(NH_3)_5H_2O]^{3+}$ and $[Rh(NH_3)_5H_2O]^{3+}$ have volumes of activation that are considerably different from those observed for $[Co(NH_3)_5H_2O]^{3+}$ and are consistent with an associative character in the mechanism.

On the whole, the large number of reactions assigned a bimolecular mechanism on the strength of certain effects in mixed solvents, unusual enthalpies of activation, volumes of activation (from the dependence of rate on pressure), and so on, are generally solvolytic and the effective criterion, that is, the ability of the reaction centre to exert a strong discrimination between different nucleophiles, either does not exist or has not been tested.

7-6 Systematic discussion of the mechanism of octahedral substitution found in relatively inert systems

The bulk of the information comes from the following systems: (a) Co(III): this is the archetype reaction system where the bulk of the early kinetic and stereochemical data was obtained. (b) Cr(III) and Rh(III): preparative difficulties (either tractability or cost) for long restricted the number of available systems and only in recent years, with all the easy stuff done elsewhere, have any but the few dedicated pioneers entered this area. (c) Ru(III) and Ir(III): interest has developed very recently in the compounds of these elements and octahedral substitution reactions are starting to present interesting patterns. (d) Ni(II), Ni(III), Pt(IV): a little information is available at the moment but it is insufficient for any overall picture to be formed.

Most of the information comes from the complexes containing mainly 'hard' ligands such as amines and lacking strongly covalently bonded ligands (one may be tolerated and generally gives rise to a strong *trans*-effect). Phosphine, arsine, and metal–carbon bonds often tend to promote redox reactions and substitution takes place by reversible oxidative addition, with or without catalysts (see Chapter 10). The reactivity of Pt(IV) complexes containing amine ligands by simple substitution is so low that these alternative mechanisms take over. With trans-$[Co(diars)_2Cl_2]^+$ exchange of chloride is catalysed by species that promote redox processes, and the rate of the apparent direct exchange decreases with the increasing purity of the compound. It is at least 10^6 times less reactive towards simple substitution than trans-$[Co\ en_2Cl_2]^+$.

With inert complexes it is convenient to systematize the discussion in terms of the reaction being studied. The solvolytic reaction and the anation reaction which will be discussed first of all are nothing more than the reverse and forward reactions of complexation that have already been discussed for labile complexes.

7-6-1 The solvolytic reaction

This is a reaction in which the solvent also plays the part of the entering nucleophile and consequently one cannot relate molecularity to rate since the concentration of the solvent remains invariant. The use of mixed solvents, especially when only one of the components is capable of coordination, is becoming a very popular area of study but while this provides considerable information about solvation it does not provide any real evidence relating to mechanism.

The process is generally reversible

$$R_5MX + S \underset{\text{anation}}{\overset{\text{solvolysis}}{\rightleftharpoons}} R_5MS + X$$

and is most conveniently studied when the equilibrium is over to the right. The position of equilibrium will depend greatly upon the nature of S, X, M, and even R. The discussion of such complex formation equilibria (instability constants) and the way in which they depend upon the various factors is beyond the scope of this book but a few points that are relevant to this discussion can be raised. For example, water is generally a good solvent for solvolysis and much of the information about the solvolytic reaction relates to aqueous solution (aquation). Most of the other oxygen donor solvents are less effective at solvolysing than water (dimethyl sulphoxide occasionally proves an exception to this rule) but liquid ammonia will solvolyse soluble cobalt(III) complexes irreversibly. The position of equilibrium is strongly dependent upon the nature of the leaving group (the ligand that is in competition with the solvent) and the central metal ion (as one might expect) and the other ligands in the complex (perhaps more surprisingly) also play an important part. When solvolysis is thermodynamically unfavourable it is possible to attempt to determine the rate of solvolysis by measuring the rate of exchange of labelled X^- with coordinated X (or even the replacement of coordinated

Table 7-5 First-order rate constants (measured at, or extrapolated to, 25°) and enthalpies of activation for the solvolytic aquation of some chloro-amine complexes

M	$[M(NH_3)_5Cl]^{2+}$		cis-$[M\ en_2Cl_2]^+$		trans-$[M\ en_2Cl_2]^+$	
	$10^7 k(s^{-1})$	$\Delta H^\ddagger$ kcal.mol^{-1}	$10^7 k(s^{-1})$	$\Delta H^\ddagger$ kcal.mol^{-1}	$10^7 k(s^{-1})$	$\Delta H^\ddagger$ kcal.mol^{-1}
Co(III)	17	23	2500	22	320	27
Cr(III)	73	24	3300	21	220	23
Ru(III)	8·0	23	450	21	—	—
Rh(III)	0·6	24	10	—	0·9	25
Ir(III)	~0·001	—	—	—	0·005	29

X by ligand Y). This of course presumes that the mechanism of ligand replacement goes by way of solvolysis and time must be spent in proving this assumption.

The invariant nature of the entering group can allow the study of the role of the following other factors.

7-6-1a Dependence of reactivity upon the nature of the central metal atom

For a series of complexes that differ only in the nature of the central atom the rate of solvolysis is markedly dependent upon the nature of M. Table 7-5 shows some rate data for the aquation of series of complexes in which the central atom is the only variable. In all cases the reactivity sequence is $Co(III)(d^6) \sim Cr(III)(d^3) > Ru(III)(d^5) > Rh(III)(d^6) \gg Ir(III)(d^6)$ and apart from occasional odd behaviour of cobalt(III) (which will be discussed later) the decrease in reactivity is paralleled by an increase in the activation energy. This sequence is usually maintained for weak *trans* effect ligands, although the relative reactivities might change considerably as the nature and position of the ligands are changed.

7-6-1b Dependence of reactivity upon the nature of the leaving group

Since the reaction is essentially dissociative, the rate of reaction will be extremely dependent upon the nature of the leaving group. For an extensive series of the type $[Co(NH_3)_5X]^{n+}$ the reactivity sequence has been shown to be HCO_3^- ($1·6 \times 10^{-3}$) $\gg NO_3^-$ ($2·6 \times 10^{-5}$) $> I^-$ ($8·3 \times 10^{-6}$) $\sim H_2O$ ($6·6 \times 10^{-6}$) $\sim Br^-$ ($6·3 \times 10^{-6}$) $> Cl^-$ ($1·6 \times 10^{-6}$) $\sim SO_4^{2-}$ ($1·2 \times 10^{-6}$) $> -SCN^-$ ($8·0 \times 10^{-7}$) $> F^-$ ($8·6 \times 10^{-8}$) $> CH_3COO^-$ ($1·6 \times 10^{-8}$) $> -NCS^-$ ($5·0 \times 10^{-10}$) $> NO_2^- > NH_3 > > OH^- > CN^-$ (last four all very slow, figures in parentheses are first-order rate constants, s^{-1}, at 25°), and similar extensive series have been built up for $[Cr(H_2O)_5X]^{n+}$ and $[Cr(NH_3)_5X]^{n+}$. The sequence is by no means immutable and depends upon the nature of the reaction centre because, in general, the aquation lability of octahedral complexes parallels their instability. Thus, the reactivity sequence for $[Co(CN)_5X]^{3-}$ is $F^- \gg Cl^- > Br^- > I^- > -SCN^-$ and the change in behaviour on replacing the 'hard' ammonia by 'soft' cyanide is what might be expected since the instability sequence for the halide complexes is reversed. It is of interest to note that, in $[Co(NH_3)_5X]^{n+}$, $X = H_2O$ is many powers of 10 more labile than $X = OH$, but in $[Co\ EDTA\ X]^{n-}$ (EDTA = N,N,N'-N'-ethylenediaminetetraacetate, here acting as a pentadentate) the reactivity difference between $X = H_2O$ and $X = OH^-$ is only a factor of three. Much the same is observed in *trans*-$[Coen_2SO_3X]^{n+}$ and $[Co(CN)_5X]^{n-}$ and this has important consequences in preparative chemistry. The relationship between lability and instability constant can be quantified as a linear free energy relationship. Thus for $[M(NH_3)_5X]^{n+} + H_2O \underset{k}{\overset{k}{\rightleftharpoons}} [M(NH_3)_5H_2O]^{(n+1)} + X^-$ the plot of log k versus log K (K is the instability constant) is linear. In the case of M = Co(III), the slope is approximately unity and this has been taken as evidence that the Co—X bond is virtually broken in the transition state (note that a truly dissociative mechanism *in solution* does not require that the M—X bond is completely broken in the transition state—the loss of X is complete in the intermediate). It is interesting to observe that the slope of the log k vs log K plot for M = Cr is only 0·6 and it has been assumed that the Cr—X bond is less extended than the Co—X bond in the transition state.

7-6-1c Effect of the other ligands in the complex

In general, the rates of substitution reactions of octahedral complexes are very sensitive to the nature of the other ligands attached to the reaction centre. In recent years it has become abundantly clear that there are at least two types of behaviour.

Type I. The rate is sensitive to the nature of the ligand but it is not very sensitive to its position with respect to the leaving group.

Type II. The rate is very sensitive to the nature of the ligand *trans* to the leaving group but not greatly affected by the *cis* ligand.

Type I behaviour is found typically in 'Werner' complexes of cobalt(III) and chromium(III) and possibly in other elements of the first row of the transition series, although it is not possible to vary independently the 'non-participating' ligands to any significant extent when studying substitution reactions of labile systems. The kinetics of the solvolysis of $[Co\ en_2A\ Cl]^{n+}$ have been studied extensively and Table 7-6 contains a selection of data. It will be seen that, apart from the case where A = —NCS, there is never more than a tenfold difference in rate between the *cis* and *trans* isomers. This may be relevant only to structurally and

Table 7-6 First-order rate constants for the aquation of cis- and trans-$[Co\ en_2ACl]^{n+}$ at 25°

A	$10^5 k(s^{-1})$ at 25°	
	cis	trans
—OH	1200	160
—Cl	24	3·5
—Br	14	4·5
—NCS	1·1	0·005
—NH_3	0·05	0·034
—OH_2	0·16	?
—CN	?	8·2
—N_3	20	22
—NO_2	11	98

mechanistically simple complexes of this sort because the difference in reactivity between cis and trans-$[Co\ cyclam\ Cl_2]^+$ is $k_{cis}/k_{trans} = 15{,}000$ at 25°. Type I behaviour can apparently be subdivided further in accordance with the electronic properties of the ligands. Those which can function as π-electron pair donors have labilizing powers that increase in the order of their π-donor ability, for example, —$NCS^- <\ CH_3COO^- <\ —Br^- \sim —Cl^- <\ —OH^- <\ —NRR'^-$ and in these cases cis labilization is stronger than trans labilization (especially in the case of —NRR'—see base hydrolysis below—and NCS) and, as will also be discussed below, substitution can take place with stereochemical change. When ligand A does not possess the ability to act as a π-donor but may form strong σ-bonds to the metal or even function as a π-acceptor, for example, —CN^-, —NO_2^-, the labilizing power is most effective from the trans position and substitution takes place with complete retention of configuration. It is possible that this is really a merging of Type I into Type II behaviour. This second type can also be found in the trans-$[Co\ en_2ACl]^{n+}$ system when A = $SO_3^=$ or NO. Such complexes are extremely labile and aquate far too rapidly for any measurement of rate constants to be possible. The ligands SO_3 and NO have a very strong trans labilizing effect and are also what has been termed 'non-innocent' by Jorgensen. This is used in the sense that one cannot readily predict the distribution of the electrons and hence the oxidation state of the metal. This is especially true of the NO complex. The trans specificity of the SO_3 group has been demonstrated in $[Co(NH_3)_5SO_3]^+$ where only one ammonia (presumably that trans to SO_3) will exchange with free $^{15}NH_3$ in the solution. The Type II behaviour is also found in cobalt(III) complexes of the type $[Co(dmg)_2AX]^{n-}$ or $[Co\ salen\ AX]^{n+}$, (salen = NN'-ethylenebissalicylaldimine), cobalamin, porphyrin, and related complexes, when A is a strong σ-donor such as —H, a C—bonded organic group, or phosphine. These strong *trans* effect ligands are precisely the ones that are found to give *trans* bond lengthening (ground state destabilization) in square planar complexes and it is thought that the same mechanism applies here. The role of the *cis* ligands is less clear. In all cases they provide at least a partially conjugated system and it may well be that they serve mainly as a means of stabilizing the Co—H, Co—C, or Co—P bonds and thereby enable the compounds to be made and studied. No one has yet been able to prepare, let alone study, the A = H, alkyl, aryl, phosphine, and so on, derivatives in the [Co en$_2$AX]$^{n+}$ series and it has not been for want of trying. There is, at present, far less systematic data available for complexes of the second and third row transition elements but in all cases the behaviour is of Type II and labilization from the *trans* ligand dominates.

7-6-2 The anation reaction

This reaction is the reverse of the solvolytic reaction and corresponds exactly to the complex formation reaction that has been discussed already for labile systems. The only reason for separating the discussion of labile systems from inert systems is that it is possible to make independent measurements of any pre-equilibrium processes, to concentrate upon a single leaving group, to examine the stereochemistry of the substitution and many other advantages besides. The discussion in this section will be restricted to the displacement of coordinated water in aqueous solution but we will meet the subject again when the non-aqueous systems are discussed. A typical reaction that has been examined in great detail is the anation of [Co(NH$_3$)$_5$H$_2$O]$^{3+}$. All of the requirements, except one, are met by this system which lacks only a suitable signpost to provide stereochemical information. (It is possible to make *trans*-[Co(NH$_3$)$_4$(^{15}NH$_3$)H$_2$O]$^{3+}$ with an isotopic signpost but it is very expensive.)

The rate of the anation process is dependent upon the concentration of the entering ligand, Y, and the I$_d$ mechanism has been assigned. Thus,

$$Co(NH_3)_5H_2O^{3+} + X^{n-} \underset{}{\overset{K}{\rightleftharpoons}} [Co(NH_3)_5H_2O^{3+} \ldots X^{n-}] \quad \text{fast}$$
$$\text{'outer sphere complex'}$$

$$[Co(NH_3)_5H_2O^{3+} \ldots X^{n-}] \overset{k_i}{\longrightarrow} Co(NH_3)_5X^{(3-n)+} + H_2O$$
$$\text{interchange}$$

If the reverse reaction is ignored and the study is made in the presence of sufficient excess of X^{n-}, the observed first-order rate constant, k_{obs}, obeys the relationship

$$k_{obs} = Kk_i[X^{n-}](1 + K[X^{n-}]).$$

In many cases $K[X^{n-}]$ remains small compared to 1 and the expression simplifies to $k_{obs} = Kk_i[X^{n-}]$. This is unfortunate because it is necessary to estimate a value for K in order to calculate the required k_i. But in favourable circumstances, notably when $X = SO_4^{2-}$, the departure from the simplified form is adequate to allow independent evaluation of K and k_i. Using either experimental or estimated values for K it has been found that k_i is essentially independent of the nature of Y. The interesting feature is that k_i is only about one-sixth of the size of the rate constant for water exchange of this complex. This is fully in accord with the I_d mechanism since it should be realized that the complex is still surrounded by water molecules when X^{n-} forms an outer sphere complex and, even though it sits there patiently waiting for the coordinated water to leave, X^{n-} is still in competition with these water molecules when it comes to entering the coordination shell. Langford suggests that this factor of $\frac{1}{6}$ represents the relative numbers, that is, that there are six suitably solvating water molecules. It is likely that the position occupied by X^{n-} and the stereochemical requirements of the substitution must also be taken into account because with the analogous $Cr(NH_3)_5H_2O^{3+}$ cation, k_i is only $\frac{1}{30}$ of the rate constant for water exchange. On the other hand, with $[Rh(NH_3)_5H_2O]^{3+}$, estimates of k_i suggest that it is significantly larger than the rate constant for water exchange. An I_a mechanism has been suggested, the idea being that X^{n-} in the outer coordination sphere has some influence on the rate at which the coordinated water leaves. The variation in k_i is not large and so the mechanism can only be just over the I_d—I_a borderline. The behaviour of Cr(III) complexes, such as $Cr(NH_3)_5H_2O^{3+}$ and $Cr(H_2O)_6^{3+}$, presents further complications in that the rates of anation are markedly sensitive to the nature of the entering group Y. (This is, of course, the reason why the slope of the plot of log k_{aq} vs log K has a slope of only 0·6.) If the interchange mechanism were assumed this would require that k_i was very sensitive to the nature of Y. In general, values of k_i calculated on the basis of plausible ion-association constants are very much less than the known rate of water exchange (for $Y = I^-$ anating $Cr(H_2O)_6^{3+}$ a reduction by a factor of 10^4 has been suggested) but on occasion they can be greater (H_3PO_2 is extremely effective but probably does not involve Cr—O fission). Thus with Cr(III) there can be a very large variation of k_i but, since it is generally much smaller than the rate of water exchange, an I_a mechanism is not mandatory unless a plausible explanation of the *very high* competing nucleophilicity of solvent water is forthcoming.

Whereas the interchange mechanism is the usual method for complex formation in labile systems and reasonably common for anation reactions of the inert complexes, it is not exclusive. One might ask what would be the consequence of making a complex with no residual charge and a non-hydrogen bonding exterior or even of taking an anionic complex and studying its reaction with anionic ligands? This has indeed been done in several cases notably $[Co(CN)_5H_2O]^{2-}$, trans-$[Co(CN)_4SO_3H_2O]^{3-}$, and $RhCl_5H_2O^{2-}$. Here the reactivity is very much greater than might be expected from an interchange relying on the pre-association of two species of like charge even though the kinetic form is similar to that of the I_d interchange. However, it arises in a different way. Thus for the truly D mechanism

$$Co(CN)_5H_2O^{2-} \underset{k_{-1}}{\overset{k_1}{\rightleftharpoons}} Co(CN)_5^{2-} + H_2O$$

$$Co(CN)_5^{2-} + X^- \overset{k_2}{\longrightarrow} Co(CN)_5X^{3-}$$

it is possible to show that

$$-d[Co(CN)_5H_2O^{2-}]/dt = \frac{k_1 k_2 [Co(CN)_5H_2O^{2-}][X^-]}{k_{-1}[H_2O] + k_2[X^-]}$$

If $[X^-]$ remains constant in any one experiment the process will be first-order and by simple rearrangement it is possible to write

$$k_{obs} = k_1 \frac{k_2}{k_{-1}[H_2O]}[X^-] \bigg/ \left(1 + \frac{k_2}{k_{-1}[H_2O]}[X^-]\right)$$

which has precisely the same form as that produced by the I_d mechanism. The two processes are nevertheless readily distinguishable. Thus, the departure from first-order dependence on $[X^-]$ in the I_d mechanism is due to the breakdown of the assumption that $K[X^-] \ll 1$. This occurs when significant quantities of the reagent are in the form of the aggregate and with any luck the presence of the aggregate can be distinguished and confirmed by non-kinetic methods. On the other hand, the deviation from simple first-order dependence on $[X^-]$ in the D mechanism comes when the assumption that $k_{-1}[H_2O] \gg k_2[X^-]$ breaks down. This occurs when more than a small fraction of the five-coordinated intermediate is snapped up by X^-, instead of taking up a water molecule and returning to the start. Thus it arises from competition and not from the build-up of any intermediate and so no non-kinetic disturbance will be observable. The second distinction is found in the magnitude of the limiting rate constant. In the I_d case we have seen that k_i can be less than the rate constant for water exchange. In the I_a mechanism it can be greater. In the D mechanism k_1 *must* be equal to the rate constant for water exchange.

It is possible to obtain values for $k_2/k_{-1}[H_2O]$ and to show that these cannot possibly be accounted for by pre-association. It does, on the other hand, make it possible to assess the discriminating power of the five-coordinate intermediate in terms of the relationship between k_2/k_{-1} and the nature of X^-. A collection of such values is given in Table 7-7 and

Table 7-7 Relative reactivities of nucleophiles with some five-coordinate intermediates (k_2/k_{-1})

Nucleophile	$Co(CN)_5^{2-}$	$Co(CN)_4SO_3^{3-}$	Co cyclam Cl^{2+}	$Co(NH_3)_4SO_3^{+}$ (a)	$RhCl_5^{2-}$	$(C_6H_5)_3C^{+}$ (b)
H_2O	1·0	1·0	1·0	—	1·0	1·0
$N_2H_5^+$	4·0	—	—	—	—	—
N_2H_4	6·3	—	—	—	—	—
NH_3	8·2	10^3	—	1·0	—	—
py	23·9	—	—	—	—	—
Cl^-	6·0	—	18	—	1·2	3100
Br^-	6·3	—	—	—	0·9	—
I^-	11·7	—	—	—	1·0	—
NCS^-	20·4	—	49,000	30	4·4	13,000
N_3^-	32·4	—	—	—	7·9	280,000
NO_2^-	—	—	—	70	5·9	—
I_3^-	422	—	—	—	—	—
CN^-	—	1900	—	43	—	—
OH^-	—	—	—	8000	—	53,000
SO_3^{2-}	—	80	—	—	—	—

(a) In aqueous solution but relative to $k_2(NH_3) = 1·0$.
(b) A relatively stable three-coordinate carbonium ion.

some interesting observations can be made. The most extensive compilation of data relates to the $Co(CN)_5^{2-}$ intermediate and it is clear that the relative reactivity can be dependent upon the nature of the nucleophile, while there is no marked correlation with charge. The extent of discrimination, which can be assessed by the magnitudes of the relative reactivities, varies considerably with the nature of the five-coordinate intermediate so that, while $RhCl_5^{2-}$ can only marginally discriminate between the range of nucleophiles presented, the other species are much more effective. It is said that there is a relationship between the discriminating power and the reactivity of the five-coordinate intermediate and it seems reasonable to believe that the greater the reactivity and the shorter the lifetime of the intermediate, the less selective it can be in its choice of bonding. Included for comparison are data relevant to the reaction of the trityl cation $(C_6H_5)_3C^+$ which is a relatively stable carbonium ion and exerts a considerable discriminating ability.

7-6-3 Base hydrolysis

This reaction is singled out for special mention because it apparently contradicts the statement that the rate of substitution of an octahedral complex is not markedly sensitive to the nature of the entering groups. This is not a general reaction but is confined mainly to a range of Co(III) amine complexes and their Ru(III) analogues. Similar complexes with Cr(III), Rh(III), and Ir(III) sometimes exhibit this reaction but the effect is far less marked.

The rate law takes the form

$$-d[\text{complex}]/dt = k[\text{complex}][OH^-]$$

and the actual rate of reaction can be very much greater than that associated with the behaviour of the complex in the absence of base, and accelerations up to and beyond a factor of 10^8 can be achieved provided the alkali concentration is high enough. The clear cut second-order kinetics and the very marked rate enhancement led some people (myself included) to postulate a simple bimolecular mechanism (A or I_a) but the lack of other bimolecular mechanisms and the accumulation of a wealth of fact in support of an alternative mechanism has caused most of us to change our minds. This alternative mechanism, while accounting for the extraordinary effect of hydroxide, allows the process to remain essentially dissociative. In it, the hydroxide functions as a base and serves to remove a proton from an amine (or ammine) ligand. This part of the mechanism is strongly supported by the fact that no complex lacking such a proton has yet been found to be markedly sensitive to base hydrolysis. The deprotonated species, called the conjugate base, is dissociatively labile and the overall process can be represented (for the reaction $[Co(NH_3)_5Cl]^{2+} + OH^- = [Co(NH_3)_5OH]^{2+} + Cl^-$) as

$$[Co(NH_3)_5Cl]^{2+} + OH^- \underset{k_{-1}}{\overset{k_1}{\rightleftharpoons}} [Co(NH_3)_4NH_2Cl]^+ + H_2O \quad (7\text{-}1)$$

$$[Co(NH_3)_4NH_2Cl]^+ \xrightarrow{k_2} [Co(NH_3)_4NH_2]^{2+} + Cl^- \quad (7\text{-}2)$$

$$[Co(NH_3)_4NH_2]^{2+} + H_2O \xrightarrow{fast} [Co(NH_3)_5OH]^{2+} \quad (7\text{-}3)$$

This mechanism is generally designated as S_N1cb, although in the Langford-Gray nomenclature adopted in this book we would say Dcb. The proton transfer reaction, Eqn (7-1) is reversible and generally, but by no means always, fast compared to the subsequent dissociation of the conjugate base. Under these circumstances it can be represented as an equilibrium with constant K. Even under these circumstances k_1 can be measured since it is statistically related to the rate constant for proton exchange. In the chloropentamminecobalt(III) complex the rate constant for proton exchange could be as much as 10^5 times greater than that for chloride release. A pre-equilibrium proton transfer would lead to specific base catalysis in aqueous solution (that is, hydroxide is the only base that is an effective catalyst) and this is in accordance with observation in most cases. For such a pre-equilibrium the full rate expression would become

$$-d[\text{complex}]/dt = \frac{k_2 K[\text{complex}][OH^-]}{1 + K[OH^-]}$$

and would reduce to the experimentally observed form if $K[OH^-] \ll 1$. This indeed holds true for nearly all cobalt(III) amine complexes where $K < 0.1$ or even < 0.01 but with coordinated aniline, which is a much more acidic amine, K for the equilibrium

$$cis\text{-}[Co\ en_2(C_6H_5NH_2)Cl]^{2+} + OH^- \rightleftharpoons cis\text{-}[Co\ en_2(C_6H_5NH)Cl]^+ + H_2O$$

is approximately $10^4 M^{-1}$ and the assumption that $K[OH^-] \ll 1$ breaks down in weakly basic solution. In the few cases where the proton transfer is of a similar rate to that of base hydrolysis, it has been shown that the act of base hydrolysis leads to the exchange of a proton and that the reaction is subject to *general* base catalysis.

Since the second-order rate constant $k = k_2K$, and we know that K is relatively small, it is necessary to account for the large magnitude of k_2, i.e. to explain why the amido ligand has such strong labilizing powers. The currently accepted explanation is that the amido group, with its spare pair of electrons functions as a π-donor and stabilizes the five-coordinate intermediate (Fig. 7-1). This is the same as the labilizing action of the less effective —OH and —Cl. This explanation is currently

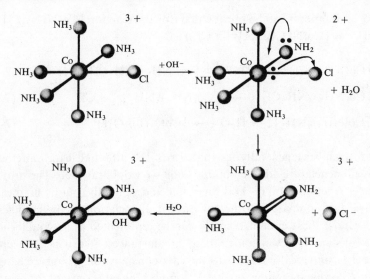

Fig. 7-1

under attack and more work will be necessary in this area. The steric course of the base hydrolysis involves considerable stereochemical change at cobalt(III) and a trigonal bipyramidal intermediate would be indicated. On the other hand, the equally labile analogous Ru(III) species react with complete retention of configuration.

The true D character of this type of reaction, that is, the existence of a labile five-coordinate intermediate has been demonstrated by classical trapping experiments. Thus, when $Co(NH_3)_5X^{2+}$ or *cis-* or *trans-*$[Co\ en_2NH_3Cl]^{2+}$ ($X = Cl, Br, NO_3$) are treated with base in the presence of a large excess of Y^- ($Y^- = NO_3^-, N_3^-, NO_2^-, NCS^-$), up to 25% of the product is in the form of $Co(NH_3)_5Y^{2+}$ or $[Co\ en_2NH_3Y]^{2+}$. The rate of reaction is unaffected by the amount of Y^- present and, while the amount of acido-product does not seem to depend upon the concentration of Y^- (over the range studied) it is very much dependent upon its nature. It is easily shown that Y^- could not have entered the coordination sphere before or after the act of base hydrolysis and so must have been trapped by a reactive, presumably five-coordinate, intermediate. It is of interest to note that when Y is NCS^- the bulk of the acido complex isolated is the unstable S-bonded isomer $[Co(NH_3)_5SCN]^{2+}$ and this constitutes at the moment the best, if not the only, method for its preparation.

7-6-4 Reactions in non-aqueous solvents

The justification for dealing with this as a separate item in no way relates to mechanism since we encounter precisely the same I_d process. However, on moving away from water to less coordinating and less solvating solvents, solvolysis is reduced sufficiently for effects that are masked by it in water to become examinable, and interactions with other solute species are enhanced. The problems of solvent competition that dominate the anation reactions of $Co(NH_3)_5H_2O^{2+}$ either vanish or are greatly reduced. Much of the work in this area has been done with complexes of the type $[Co\ en_2 AX]^{n+}$ in alcohols and dipolar aprotic solvents such as dimethylsulphoxide, dimethylformamide, dimethylacetamide, acetone, tetramethylenesulphone, and so on. The solvolytic interference can be considerable, as in the case of dimethylsulphoxide, or negligible, as in the case of tetramethylenesulphone, but it is often relatively easy to distinguish direct substitution from cryptosolvolysis. (Cryptosolvolysis is a process whereby one ligand becomes replaced by another by means of a solvolysis followed by an anation under conditions where the solvento complex is too reactive to be directly observable, that is,

$$R_5MX + S \rightleftharpoons R_5MS + X \xrightarrow{+Y} R_5MY + S)$$

The rates of the substitution reaction are often dependent upon the concentration of the entering nucleophile and this dependence can be quite complicated. Fig. 7-2 shows a number of curves for the plot of k_{obs} against $[Y]$ which have been obtained. Apart from one or two odd forms (not shown) which can be attributed to experimental incompetence, the rest can be accounted for in terms of a series of pre-equilibria, each generating an aggregate with its own rate constant for interchange.

$$R_5MS \ldots S \underset{k_1}{\overset{+Y}{\rightleftharpoons}} R_5MX \ldots Y \underset{k_2}{\overset{+Y}{\rightleftharpoons}} R_5MX \ldots 2Y \rightleftharpoons \text{and so on}$$

$$\downarrow k_0 \qquad \qquad \downarrow k_1 \qquad \qquad \downarrow k_2$$

$$R_5MS \qquad \qquad R_5MY \ldots X \qquad R_5MY \ldots X \ldots Y$$

$$(R_5MY) \qquad (R_5MS \ldots X \ldots Y) \quad (R_5MS \ldots X \ldots 2Y)$$

This would give rise to a general expression of the form:

$$k_{obs} = \frac{k_0 + k_1 K_1[Y] + k_2 K_1 K_2[Y]^2 \ldots}{1 + K_1[Y] + K_1 K_2[Y^2] \ldots}$$

and suitable relationships between the individual rate and equilibrium constants can account for all the observed dependences. In the case of charged complexes and reagents it has been possible to measure the pre-equilibrium constants independently by conductimetric and spectrophotometric techniques. Quite often these association constants are very large and considerably in excess of anything calculated on the basis of a simple ion-association model in which ionic charge and size and solvent dielectric constants are the only variables fed into the calculation. A

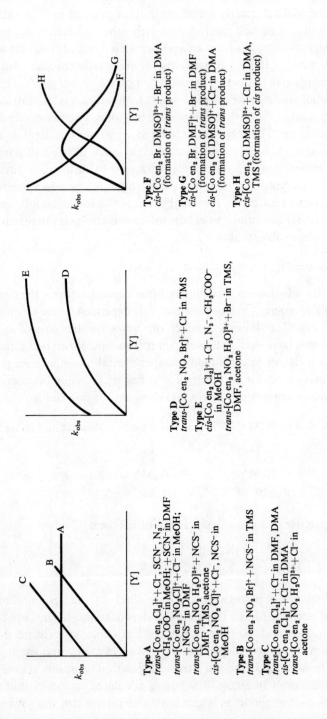

Fig. 7 – 2 Dependence of pseudo first-order rate constant k_{obs} for entry of Y upon the concentration

major contribution appears to come from the amine protons since replacement of these by methyl groups reduces the ion-association to levels acceptable to the simple electrostatic model. When the comparison has been made, the kinetic and non-kinetic values of K have been in reasonably close agreement. The possible formation of ion-aggregates of higher complexity than the ion-pair is well established and the interesting and perhaps surprising observation is that, in certain circumstances, the cationic charge can be overcompensated, for example, $trans$-$[Co\ en_2NO_2H_2O]^{2+}$ will associate with a maximum of $three$ bromide ions in acetone at 25°.

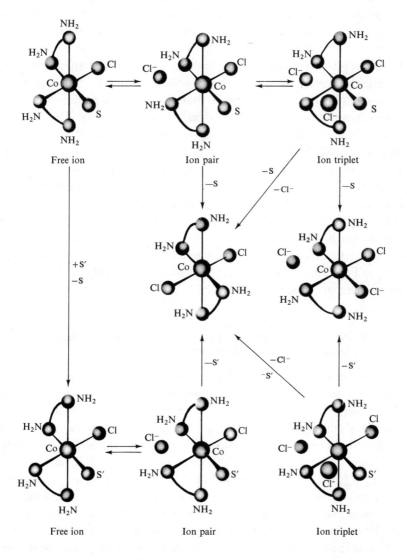

Fig. 7 – 3 Chloride anation of cis-$[Co\ en_2\ S\ Cl]^{2+}$ in solvent S'

Analysis allows the various interchange constants to be estimated and while the story is reasonably complicated and a great deal of work remains to be done, the reaction of *cis*-[Co en$_2$DMSO Cl]$^{2+}$ (DMSO = dimethylsulphoxide) with chloride in dimethyl formamide is reasonably typical and presents a system that has been investigated in some detail. Here it can be shown, under non-reacting conditions, that on raising the chloride concentrations in the solution the complex is distributed first between the free ion and the chloride ion-pair and at higher chloride between the ion-pair and the ion-triplet. The rate of formation of the dichloro complex increases with chloride ion concentration as the substrate accumulates as the ion-pair but further increase in chloride ion concentration, while increasing the concentration of the ion-triplet at the expense of the ion-pair, does not significantly affect the rate. On the other hand, the product changes. At the lower chloride concentrations the main product is the *trans*-dichloro complex while at higher chloride concentrations it is the *cis*-dichloro isomer that is formed in preference. These observations have been interpreted in the following way. The free ion cannot undergo chloride anation and only experiences solvolysis (replacement of DMSO by DMF). The ion-pair and ion-triplet undergo anation at much the same rate since in both cases the chloride has to wait for the DMSO to dissociate before it can enter the complex. However, the ion-pair gives mainly *trans*-product since the chloride ion associates with the complex as far as possible from the coordinated chloride, while the ion-triplet, which must now have a chloride adjacent to the leaving DMSO, can give either *cis* or *trans* dichloro product. This is shown diagrammatically in Fig. 7-3, and is one of the few examples currently available of the stereochemistry of ion-association.

7-7 A unified view of the dissociative, D, and dissociative interchange, I$_d$, mechanism

The general features of a substitution reaction with dissociative activation are shown in Fig. 7-4. Here the circles around the complexes represent their inner solvation shells and, when unbroken, indicate that these are composed entirely of solvent molecules. Other solute species, represented by X and Y (which may be neutral, anionic, or even cationic), compete with the solvent for places in the inner solvation shell. When X or Y occupy a place in the inner solvation shell this is represented by making them part of the ring. Although only a single entry is represented there is no reason to stop at this point and, when Y is a neutral species, this picture can be used to represent differential solvation in the mixed solvent Y + S where, in the limit, the solvation shell contains only Y molecules. When the substrate is cationic and Y anionic, the process represents ion-association.

Sec. 7-7 Dissociative and dissociative interchange mechanism

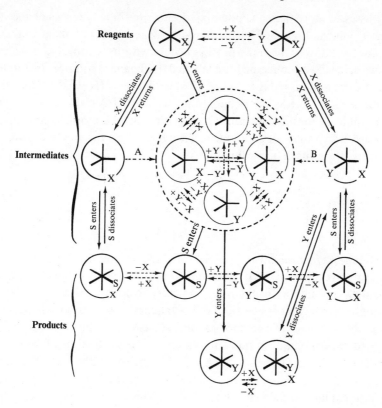

Fig. 7 – 4 The relationship between the I_d and the D mechanism. The symbolism is explained in the text

These solvation shell processes are diffusion controlled and are represented in Fig. 7-4 by dashed arrows. This type of pre-equilibrium represents the first relaxation process observed in complex formation. The first process involving the coordination shell is the rate controlling dissociation of X which, being dissociatively activated, ought to be essentially independent of the nature of Y. A small difference in the dissociation rate constant for the free substrate on the left and the aggregate on the right might be acceptable as a 'solvation effect'. The distinction between an I_d and a D mechanism lies in the fate of the species thus generated and it is necessary to consider separately the intermediate generated by the free complex, which is on the left-hand side of the figure and that generated by the aggregate, which is on the right. Three paths are available for the first type of intermediate namely: (i) return of X, (ii) entry of solvent, and (iii) equilibration of its inner solvation shell (represented by path A).

Return of X will cancel out the original act of dissociation and usually will be unobservable directly. However, if X is an ambident ligand—that is, possesses more than one potential donor site—it may have a chance to turn around and reattach itself differently. Many cases of linkage isomerization are known where the process occurs without significant exchange between the ligand concerned and labelled free ligand in solution. Examples are known for X = NCS, NO_2, CN, and even $^{15}N\equiv{}^{14}N$. Thus the unstable $Co(NH_3)_5SCN^{2+}$ changes to the stable $[Co(NH_3)_5NCS]^{2+}$ in acid aqueous solution without any parallel solvolysis or exchange with labelled thiocyanate in solution. In base hydrolysis the amido conjugate base partitions itself according to

$$[Co(NH_3)_4NH_2SCN]^+ \begin{array}{c} \nearrow [Co(NH_3)_4NH_2NCS]^+ \quad 26\% \\ \underset{+H_2O}{\overset{-SCN}{\searrow}} [Co(NH_3)_5OH]^{2+} \quad 74\% \end{array}$$

The alternative mechanism invoked for these 'intramolecular' isomerizations involves either a π-type interaction or else a temporary increase in the coordination number of the reaction centre, neither process fitting comfortably within what is known about octahedral substitution. The second fate, whereby one of the surrounding solvent molecules slips into the created vacancy requires no commitment about the lifetime of the intermediate and can take place while X is in the inner solvation shell. The third fate, on the other hand, requires that the intermediate is sufficiently long lived to equilibrate its own solvation shell. This is represented in Fig. 7-4 by the processes within the large dashed circle. The equilibrated intermediate will be able to react with solvent or with Y or even with X, but in the last case, there will have been exchange with free X in solution. In sophisticated studies it might be possible to recognize intermediate regions in which there is only time for partial equilibration of the solvation shell. The reaction of the fully equilibrated intermediate will be independent of its past history (whether it formed by path A or path B) and the nature of the leaving group X and a measure of discrimination between potential reagents, Y, will be exercised. In general, the discrimination will increase with the lifetime of the intermediate.

The intermediate derived from the aggregate will have four possible paths available. The first three are equivalent to those discussed for the left-hand intermediate but in addition there is the possibility that the Y now present and eagerly waiting can slip into the vacancy created by loss of X.

The distinction between the D and the I_d mechanism thus becomes clear. If the five-coordinate intermediate can live long enough to equilibrate its solvation environment we find all the characteristics of a classical D mechanism. The discrimination power of the intermediate is strong and independent of the nature of the leaving group, the steric course is independent of the nature of the leaving group and the existence or non-existence of pre-association is irrelevant to subsequent behaviour. If, on the other hand, the intermediate does not live long enough to equilibrate its solvation shell, the mechanism becomes I_d and the only way for a ligand Y to enter the complex is for it to take its place in the solvation shell before bond dissociation takes place and then hope that it can beat the solvent molecule when a vacancy becomes available.

This picture closely parallels that discussed for the dissociative reactions of aliphatic carbon compounds in terms of intimate and solvent separated ion-pairs and the ability to trap the carbonium ion at different stages of its development. Thus, optically active 4,4-dimethylbenzhydryl thiocyanate undergoes linkage isomerization, racemization, thiocyanate exchange, and solvolysis at different rates because each represents a trapping of the carbonium ion at different stages of its development.

7-8 Stereochemistry of octahedral substitution

The occurrence of stereochemical change, either slowly or rapidly, is extremely common in octahedral systems but if we examine the consequences of substitution we find that in all but a narrow range of compound types (of which there are a disproportionately large number of examples) the normal octahedral substitution takes place with complete retention of configuration. The evidence strongly suggests that the normal five-coordinate intermediate retains a tetragonal pyramidal arrangement.

The reactions in which it can be demonstrated that the act of substitution is actually accompanied by stereochemical change are limited to the following:

(a) Aquation of *trans*-$[Co\ L_4AX]^{n+}$ where L_4 represents four donor nitrogens, which may be monodentate, as in *trans*-$[Co(NH_3)_4Cl_2]^+$, bidentate, as in *trans*-$[Co\ en_2Cl_2]^+$, or quadridentate, as in *trans*-$[Co\ trien\ Cl_2]^+$ and A = —OH, —F, —Cl, —Br, —NCS, —OCOR. Even here the change can be prevented by a suitable choice of ligand for L_4 although the actual rate of loss of X is not greatly affected. There is some suggestion that certain analogous Cr(III) complexes give some *cis* product but here L_4 is almost as easily displaced as X and many ambiguities exist. Table 7-8 contains a set of representative data and shows how the L_4 group can also prevent stereochemical change. The *cis* isomers all aquate with complete retention of configuration.

(b) Non-solvolytic reactions of $[Co\ L_4AX]^{n+}$. Here the range of compounds examined is much more restricted but steric change is observed when A = Cl, Br, and possibly OH and is found for *cis* as well as *trans* isomers.

(c) Base hydrolysis of $[Co\ L_4AX]^{n+}$. Here, a very wide range of

Table 7-8 Steric course of aquation of some trans-$[CoL_4AX]^{n+}$ cations

L_4	A	X	Product % cis	% trans
$(NH_3)_4$	Cl	Cl	55	45
en_2	OH	Cl	75	25
en_2	Cl	Cl	35	65
en_2	Br	Br	30	70
en_2	NCS	Br	45	55
en_2	CH_3COO	Cl	75	25
[a] SS-trien	Cl	Cl	100 (β-cis)	0
RR-2,3,2-tet	Cl	Cl	50 (β-cis)	50
RS-2,3,2-tet	Cl	Cl	0	100
SS-3,2,3-tet	Cl	Cl	0	100
cyclam	OH	Cl	0	100
cyclam	Cl	Cl	0	100
tet a	Cl	Cl	0	100
tet b	Cl	Cl	0	100
RS-trans[14]diene	Cl	Cl	0	100
SS-trans[14]diene	Cl	Cl	0	100

[a] The labels R and S refer to the absolute configuration of the secondary nitrogen atoms when coordinated.
trien = $NH_2CH_2CH_2NHCH_2CH_2NHCH_2CH_2NH_2$
2,3,2-tet = $NH_2CH_2CH_2NHCH_2CH_2CH_2NHCH_2CH_2NH_2$
3,2,3-tet = $NH_2CH_2CH_2CH_2NHCH_2CH_2NHCH_2CH_2CH_2NH_2$
Cyclam, tet a, tet b, and trans[14]diene are a group of macrocyclic quadridentate nitrogen donor ligands.

ligands A will lead to some steric change, irrespective of whether they are cis or trans to the leaving group. In all of these base catalysed reactions the reactive species is the amido conjugate base and the ligand —NRR', produced by deprotonation of L, is one more ligand to add to the six listed in (a) above.

The occurrence of stereochemical change can be understood in terms of a trigonal bipyramidal geometry for the five-coordinate intermediate (Fig. 7-5). If one assumes that the loss of X is accompanied by the movement of a pair of mutually trans ligands, each cis to the leaving group, each octahedron can generate two trigonal bipyramids. If this path is reserved for recombination it corresponds to entry at a trigonal edge. Since there are three such edges in each of two trigonal bipyramids there are six possible paths, four of which lead to a rearrangement in the relative positions of the ligands in the complex.

Sec. 7-8 Stereochemistry of octahedral substitution 105

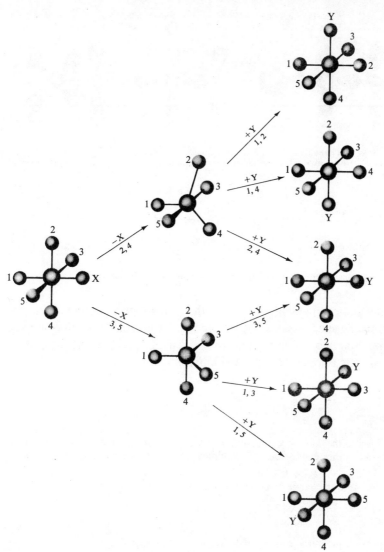

Fig. 7 – 5 Substitution leading to stereochemical change—the trigonal bipyramidal intermediate

A common feature of all the ligands that lead to stereochemical change is the presence of a pair of electrons in an orbital of the right symmetry to allow ligand to metal π-bonding. It has been suggested that this type of donation is most effective when ligand A is in the trigonal plane of the five-coordinate intermediate (Fig. 7-6). This type of interaction has already been discussed in Section 7-6-1(c) since it is the cause of the labilizing power of these ligands. The effect is mainly one of transition state stabilization, the energy of the intermediate and indirectly that of

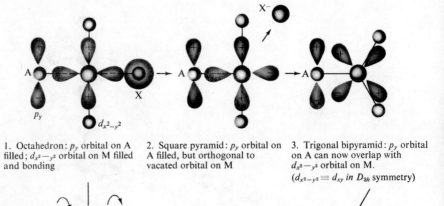

Fig. 7-6 π-donation from A to stabilize a trigonal bipyramidal intermediate

the transition state being minimized through adopting a trigonal bipyramidal arrangement. It is also clear from Fig. 7-6 that the distinction between *cis* and *trans* positions relative to the leaving group in an octahedral ground state is lost in the trigonal bipyramidal intermediate.

Problems

7-1 Derive the kinetic expression for the dissociative anation reaction. How would the expression be modified if the dissociation of the product $R_5MX \xrightarrow{k_{-2}} R_5M + X$ had to be taken into account?

7-2 For an interchange reaction obeying the kinetic relationship

$$k_{obs} = \frac{k_o + k_1 K[Y]}{1 + K[Y]},$$

what is the initial slope of the plot of k_{obs} versus $[Y]$?
[Hint: It is not $k_1 K$!]

7-3 It has been suggested that the rate determining step in the base hydrolysis of *trans*-[Co cyclam Cl_2]$^+$ is the removal of an amine proton to form the conjugate base. What effect would this have on the study of the kinetics in buffer solution?

7-4 [Co(NH$_3$)$_4$(CH$_3$NHCH$_2$COO)]$^{2+}$ (the N-methylglycine ligand acting as a chelate through N and O) has been resolved into optically active forms. From where does the asymmetry arise? Racemization and exchange of the secondary amine proton follow the same rate law (rate = k[complex][OH$^-$]) but k for exchange is some 4000 times greater than k for racemization. What do you conclude from these observations? *See*: Sargeson, A. M., B. Halpern, and K. R. Turnbull, *J. Amer. Chem. Soc.*, 1966, **88**, 4360.

7-5 The preparation of *trans*-[Co(NH$_3$)$_4$(^{14}NH$_3$)Cl]$^{2+}$ starts with Co(NH$_3$)$_5$SO$_3^+$. Explain why this is a suitable starting material and write down the steps of the synthesis.

7-6 It has been shown that the appearance of $Cr(H_2O)_5Cl^{2+}$ during the aquation of $Cr(H_2O)_5I^{2+}$ in the presence of chloride cannot be explained by anation of a $Cr(H_2O)_6^{3+}$ intermediate. It has also been shown that aquation of $Cr(H_2O)_5I^{2+}$ in ^{18}O labelled water leads to $Cr(H_2O)_6^{3+}$ with *two* labelled water molecules. Is this compelling evidence for a D mechanism? Discuss. *See*: Moore, P., F. Basolo, and R. G. Pearson, *Inorg. Chem.*, 1966, **5**, 223.

Bibliography

Kustin, K., Fast metal complex reactions, *Prog. Inorg. Chem.* (Ed. J. O. Edwards), 1970, **13**, 107.

McAuley A. and J. Hill, Kinetics of metal ion complex formation in solution, *Quart. Revs.*, 1969, **23**, 18.

Pratt, J. M. and R. G. Thorpe, *Cis* and *trans* effects in cobalt(III) complexes, *Adv. Inorg. Chem. Radiochem.* (Ed. H. J. Emeléus and A. G. Sharpe), 1969, **12**, 375.

Tobe, M. L., The role of ion-association in the substitution reactions of octahedral complexes in non-aqueous solution. *Advances in Chemistry Series*, No. 49, p. 7, A.C.S., 1965.
This volume which contains ten review papers, gives considerable insight into the background of a wide range of mechanistic problems that were current in the mid sixties. The discussions are well worth reading.

Tobe, M. L., Base hydrolysis of octahedral complexes, *Acc. Chem. Res.*, 1970, **3**, 377.

Wilkins, R. G. and M. Eigen, The kinetics and mechanism of formation of metal complexes. *Advances in Chemistry Series*, No. 49, p. 55, A.C.S., 1965.

Wilkins, R. G., Mechanisms of ligand replacement in octahedral nickel(II) complexes, *Acc. Chem. Res.*, 1970, **3**, 408.

8 Stereochemical change

8-1 Introduction

Until recent years any study of stereochemical change required a reasonably slowly reacting system in which the coordination shell was adequately signposted to allow changes in the relative positions of the ligands within it to be observed. In other words it was necessary to separate stereoisomeric forms and study their interconversion. This approach included both geometric isomers and optical isomers and Ingold showed in the early fifties that, for a complex such as $D\text{-}cis\text{-}[Co\ en_2Cl_2]^+$, $cis \rightleftharpoons trans$ and $D \rightleftharpoons L$ changes were stereochemically equivalent and differed only in the relationship between the signpost and the migrating group. With more elaborate signposting (generally requiring more elaborate means of investigation) a larger amount of stereochemical information could be obtained.

In recent years, a new and extremely powerful tool—nuclear magnetic resonance—has been applied to these problems. Not only does it provide a means of identifying and analysing isomeric forms in the course of the reaction, it also allows the study of systems at equilibrium where the reversible stereochemical changes are occurring rapidly. Measurements made over a range of temperatures have indicated and explained phenomena whose existence was not recognized before the measurements were made. The information gained from these studies will be a major part of this chapter.

A useful subdivision of the various types of stereochemical change can be made along the following lines:

8-1-1 Stereochemical change that can be proved to be a direct result of ligand substitution

A typical example here is the reaction, $trans\text{-}[Co\ en_2Cl_2] + H_2O \rightarrow 30\%\ cis + 70\%\ trans\text{-}[Co\ en_2H_2OCl]^{2+} + Cl^-$, where it can be shown that the 30% cis product is formed as a direct consequence of the act of aquation and not by any prior or subsequent act of isomerization. Such changes are the natural consequences of the geometry of the transition states and intermediates of substitution. These processes have been adequately discussed in the chapters on substitution and need not be repeated here.

8-1-2 Stereochemical change that is probably but not proved to be the result of substitution

A good example here is to be found in a range of aquo complexes of the type cis and $trans\text{-}[Co\ en_2AH_2O]^{n+}$ which isomerize and racemize readily

Sec. 8–1 Introduction 109

in aqueous solution. In the cases where A = H₂O and NCS, studies of water exchange using $H_2^{18}O$ suggest that the isomerization is indeed the consequence of the substitution of water for water occurring with stereochemical change. The situation is less definite when A = NH₃ since water exchange is considerably faster than isomerization and when A = NO₂ the story must surely be much more complicated because the rate of isomerization is independent of pH even over the range where [Co en₂NO₂H₂O]²⁺ changes to [Co en₂NO₂OH]⁺. These reactions require a great deal more study and may prove to be of considerable interest but at the moment there is inadequate information to warrant further discussion.

8-1-3 Stereochemical change without substitution but with possible bond breaking

A very large number of systems have been studied in which stereochemical change can be observed under conditions where none of the ligands exchange with similar species in the environment. Such processes would truly warrant the description 'intramolecular'. Examples of this behaviour are to be found in the reactions of *cis*-[Co(diars)₂Cl₂]⁺ which, in acidified methanol, changes to the *trans* form without exchanging its chloride;

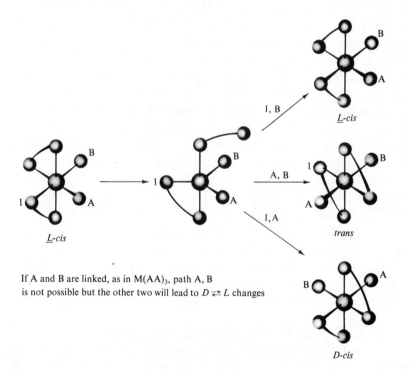

If A and B are linked, as in M(AA)₃, path A, B is not possible but the other two will lead to $D \rightleftarrows L$ changes

Fig. 8 – 1 Isomerization of M(AA)₂ AB by one-ended dissociation through a trigonal bipyramidal intermediate. (One of a number of possible paths)

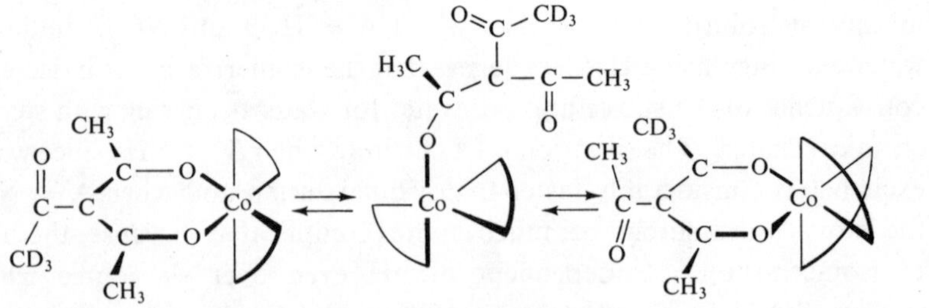

Fig. 8 – 2 'Exchange' of CD_3 with CH_3 in tris(acetylacetylacetonato) cobalt (III) as a result of ring opening. (The racemization also indicated is not mandatory).

trans-$[Co\ en_2OHNH_3]^{2+}$, which changes to the cis form without exchanging OH or NH_3; cis and trans-$[Co\ en_2(OH)_2]^+$ which racemize and isomerize with only fractional exchange of labelled OH; and so on. These examples are taken from a very wide range of octahedral complexes that behave in this way all of which have one thing in common, namely the presence of at least one chelate ring which can open to generate a temporarily five-coordinate species. Ring closure in a different position will result in stereochemical change (Fig. 8-1). Wherever this has been studied in detail there has been no evidence for a temporary solvolysis to maintain the coordination number of the original substrate. Thus, if the isomerizations of $[Co\ en_2NH_3OH]^{2+}$ or $[Co\ en_2(OH)_2]^+$ had involved opening of an ethylenediamine ring and the subsequent entry of a water molecule, we would have expected the oxygens to exchange, at least in part, with those of solvent water. Since this was not observed it must be concluded that solvolysis does not occur. A much greater collection of data comes from the studies of tris-chelated complexes, where the dissymetric octahedral species can be resolved and their racemization studied. This was a very popular pastime some years ago but is now replaced by n.m.r. studies of similar systems which do not need resolution. A very elegant study that showed the existence of ring opening was made some time ago by Piper who acetylated $[Co(acac)_3]$ with CD_3CO^+. Ring opening would allow the CH_3 and CD_3 groups to exchange their positions (Fig. 8-2) and the observation that the rate of this exchange (measured by n.m.r.) and the rate of racemization were very similar suggested most strongly that a ring opening mechanism similar to that shown in Fig. 8-1 was causing the racemization. It is likely, but by no means proved, that many racemization reactions of tris-chelated octahedral complexes go by way of ring opening.

8-1-4 Stereochemical change without any bond fission

It is reasonable to claim that reactions in this category have become one of the most exciting areas in reaction mechanisms in the past decade. In general terms, the demonstration of this phenomenon had to wait upon the

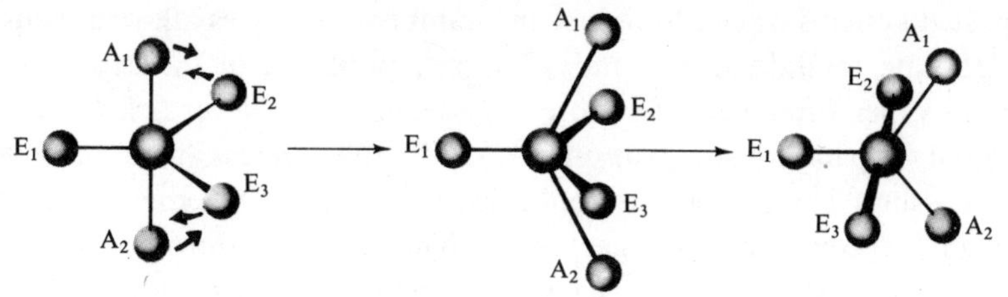

Fig. 8 – 3 'Berry' mechanism for pseudorotation in a five-coordinate species. Ligand E_1 acts as pivot for this rotation

advent of a suitable tool and nuclear magnetic resonance proved ideal for this purpose.

Early indications that certain compounds were non-rigid in this sense came from studies of the ^{19}F n.m.r. spectrum of PF_5 and the ^{17}O n.m.r. spectrum of $Fe(CO)_5$. Instead of two peaks corresponding to the two magnetically inequivalent nuclei (axial and equatorial) which would be expected for the known trigonal bipyramidal geometry, only one was observed. Accidental coincidence was ruled out as more and more examples of five-coordinate compounds behaving in this way came to light. It is known from isotopically labelled ^{14}CO exchange that $Fe(CO)_5$ holds on to its ligands tightly and the splitting of the resonance due to $^{19}F-^{31}P$ coupling in PF_5 indicated clearly that the equivalence could not be due to rapid ligand dissociation and recombination. It was proposed therefore that there was some means whereby axial and equatorial ligands could change places without breaking (or even significantly stretching) the metal–ligand bond. A simple path to allow this is shown in Fig. 8-3. In this particular case the transition state (or intermediate) has a square-pyramidal geometry. This process has been termed 'pseudorotation' and is only one of a number of pathways that can lead to a rearrangement of the coordination shell in a five-coordinate system.

8-2 General consideration of pseudorotation and the consequent topological changes

It will soon become clear that it is not too difficult (hot coffee and cold towels might be needed) to formulate and systematize all possible topological changes in terms of networks and transition states. This is true if we restrict our research to pencil and paper but the translation to real chemical systems is extremely difficult. The number of complexes where N different ligands occupy the N different coordination sites is rare indeed above $N = 4$ and so it is necessary either to work with inadequately signposted systems or else to reduce the number of changes allowed by using elaborate multidentate ligands. The interpretation of the experimental results then often becomes more or less equivocal. Nevertheless, this is a rapidly expanding area that offers considerable interest and is well worth discussing. The problem is conveniently discussed in terms of what we might call complementary geometries. The act of pseudorotation requires that certain bond angles in the parent compound change until the coordination shell assumes its complementary geometry in what would constitute either the transition state or the intermediate. Further motion regains the original geometry, but may very well lead to a change in the relative positions of the individual ligands. The concept of complementary geometries is introduced because in appropriate situations the geometries can be reversed so that, in the same way as the trigonal bipyramid can undergo pseudorotation through a square pyramidal transition state, a square pyramid can isomerize through a trigonal bipyramid transition state.

The ease with which these processes can occur will depend upon the energy difference between the complementary geometries (the ground state and the transition state) which constitutes the barrier to the process. When the complementary geometries are energetically close, either because of the nature of the ligands or because of the inherent properties of certain high coordination numbers, the barrier to pseudorotation will be small. In many of the cases which will be considered below the barrier will be increased by the need for the system to traverse an intermediate in which ligand distributions or unfavourable bond angles lead to instability. The best example of pseudorotation and stereochemical non-rigidity will be found in those areas where the barrier between complementary geometries is significantly smaller than the free energy of activation for bond fission.

It is convenient to subdivide the discussion according to the various coordination numbers encountered.

8-2-1 Inversion in three-coordinate systems

A pyramidal arrangement of three different ligands about a three-coordinate centre can be made in two enantiomeric ways which are interconvertible by way of the complementary trigonal planar form which has a plane of symmetry. This process can be likened to the 'turning inside out' of the pyramid (Fig. 8-4). Many examples of this type of inversion are known and the kinetics have been studied in detail. One of the first to be recognized was the inversion of the ammonia molecule which is best represented as a vibrational mode of this molecule with the hydrogen atoms 'tunnelling' from one position to another rather than going through a planar NH_3 transition state. One can find many more acceptable examples in nitrogen chemistry although until recently the nitrogen

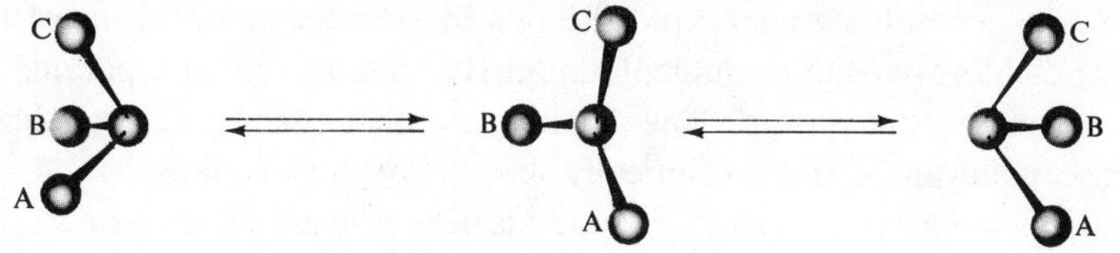

Fig. 8 – 4 Inversion of a pyramidal three-coordinate complex by way of a trigonal planar transition state or intermediate

generally had to be part of a tight three- or four-membered ring for the inversion to be slow enough to measure. Thus, rate constants in the region of 10^5 s^{-1} at 25° could be measured for nitrogen inversion in aziridines. The inversion of dibenzylmethylamine becomes slow enough to observe by n.m.r. line broadening at $-135°$. On changing from nitrogen to phosphorus, the situation changes dramatically and the barrier to inversion increases from ~ 5 kcal.mol^{-1} to > 20 kcal.mol. This means that the rate of inversion can be quite slow at room temperature and it is possible to separate the enantiomers, and study their racemization by classical methods at elevated temperatures. The resolution is generally made using the four-coordinate tetrahedral phosphorus compound which is then suitably treated to form the three-coordinate species under stereoretentive conditions. An example is to be found in the resolution of a tetrahedral phosphine oxide $R_1R_2R_3P=O$ which is then reduced stereospecifically to the phosphine by Si_2Cl_6. Alternatively, the phosphine can be coordinated to a metal, the complex resolved, and the ligand then displaced by another. For example,

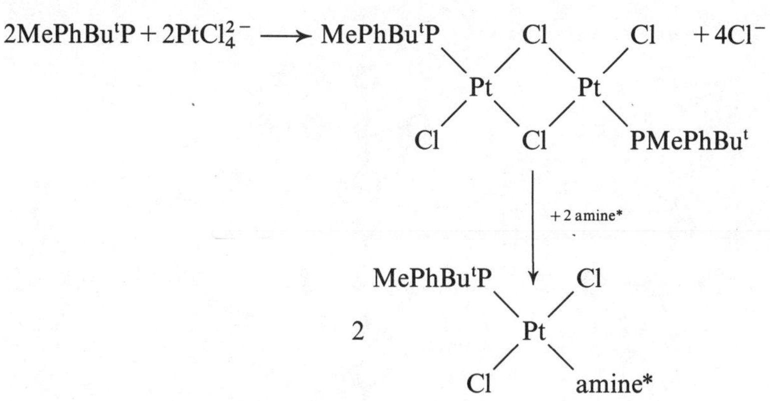

$$2\text{MePhBu}^t\text{P} + 2\text{PtCl}_4^{2-} \longrightarrow \text{MePhBu}^t\text{P}\diagdown\text{Pt}\diagup^{\text{Cl}}\diagdown\text{Pt}\diagup^{\text{Cl}}\diagdown\text{PMePhBu}^t + 4\text{Cl}^-$$

$$\Big\downarrow +2\,\text{amine*}$$

$$2\;\; \text{MePhBu}^t\text{P}\diagdown\text{Pt}\diagup^{\text{Cl}}\diagdown\text{amine*}$$

The optically active amine,

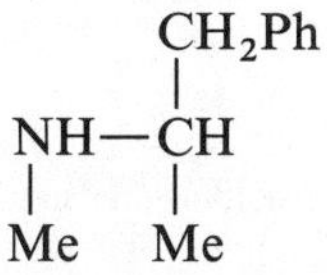

$$\begin{array}{c}\quad\quad\;\;\text{CH}_2\text{Ph}\\ \quad\quad\;\;\;\;\;|\\ \text{NH}-\text{CH}\\ |\quad\quad\;\;|\\ \text{Me}\;\;\;\text{Me}\end{array}$$

serves as a resolving agent and the two diastereoisomeric forms of the *trans* dichlorophosphineamineplatinum(II) complex can be separated by fractional crystallization. The ligands are then displaced by adding potassium cyanide and the optically active phosphine isolated.

Three-coordinate arsenic compounds behave in a similar way but have not been studied as extensively.

8-2-2 Inversion and pseudorotation in four-coordinate systems

The two complementary geometries to be considered here are the planar and the tetrahedral arrangements (Fig. 8-5). In most areas of the periodic table, the energy difference between the tetrahedral and planar forms of four-coordination is extremely large. It has been estimated that the barrier to inversion of a tetrahedral carbon atom (that is, the energy difference between the tetrahedral and planar forms) is in the region of 100-200 k cal.mol^{-1} in the types of compound one usually meets. Thus, it will always be easier to break a bond than to invert the molecule. This has not stopped people from trying to work out what properties should be possessed by the ligands likely to stabilize a planar four-coordinate carbon atom. The same story applies for all of the other P-block compounds where, although the barriers to pseudorotation may be less, the resistance to bond fission is also less.

The tetrahedral and square planar geometries approach one another in energy in the four-coordinate complexes of the elements of the first-row transition series. As was mentioned in Chapter 5, the tetrahedral and four-coordinate planar complexes of Ni(II) of the type NiL$_2$X$_2$ (where L

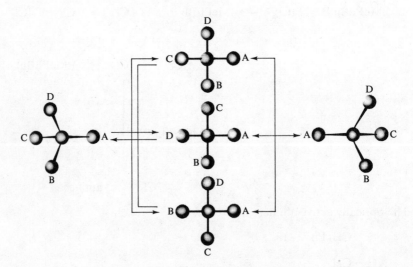

Fig. 8-5 Tetrahedral inversion by way of the complementary square planar transition state. The planar complexes can isomerize through a tetrahedral transition state

is a phosphine or arsine and X is a halide) approach one another so closely in energy that isomeric forms can be isolated. Thus, whereas nickel complexes with aliphatic phosphines, for example, $[Ni(PEt_3)_2Cl_2]$, are generally diamagnetic and square planar, those with the aromatic phosphines, for example, $Ni(PPh_3)_2Br_2$, are paramagnetic and tetrahedral. At the borderline, as for example in the case of $Ni(PPh_2Et)_2Br_2$, both forms are known and in one extreme case, $[Ni(BzPh_2P)_2Br_2]$, both geometries are found in the same unit cell of the crystal. Such isomeric pairs have been termed 'allogons' and the equilibria between them have been studied extensively. The rates of interconversion are generally very fast.

Away from this borderline region, the unstable geometry can constitute a suitable path for the *cis* ⇌ *trans* isomerization of planar complexes or the inversion of tetrahedral complexes.

On going to the corresponding elements of the second and third transition series, Pd and Pt, nature comes down strongly in favour of the planar arrangement and the barrier to pseudorotation becomes impassable. The possibility of a tetrahedral excited state in a photochemical isomerization process is still quite likely.

8-2-3 Pseudorotation in five-coordinate systems

As has already been pointed out in this and previous chapters, the complementary geometries of five-coordination are energetically close to one another. This is not only true for the regular geometries, that is, trigonal bipyramidal and tetragonal pyramidal, it also applies to the distorted structures that lie between these two. As a consequence, the tendency for five-coordinate systems to undergo pseudorotation is quite considerable. Indeed, all of the ML_5 molecules that have been studied by n.m.r. have all five L ligands magnetically equivalent even at the lowest temperatures examined. This indicates that the barrier to pseudorotation cannot exceed a few kcals per mol. It is possible to raise the barrier sufficiently to slow down this process or even to prevent it and this can be done, in at least two ways, by a suitable choice of ligands which interpose unfavourable trigonal bipyramidal intermediates in the interconversion pathway. The first way keeps five monodentate ligands but varies their electronegativity and can be illustrated by the behaviour of a sequence of compounds derived from PF_5. The monosubstituted compounds LPF_4 as a rule are as stereochemically labile as the PF_5 parent and the four fluorines remain magnetically equivalent down to the lowest temperatures. Two of the few exceptions to this rule are $(CH_3)_2NPF_4$ and $RSPF_4$ (R is an alkyl or an aryl group) where the low temperature n.m.r. shows, first, a line broadening and at the lowest temperatures, different signals from the axial and equatorial fluorines. The Berry twist (Fig. 8-3), named after its proposer, can account for the observations since the less electronegative substituent can always act as a pivot and need never compete for the axial positions favoured by the electronegative fluorine. This is not possible in compounds of the type L_2PF_3 where a Berry twist must make at least one L go to an axial position and compounds of this type are generally far more rigid, especially where L is a great deal less electronegative than fluorine. Thus $(CH_3)_2PF_3$ has both methyl groups in the trigonal plane and even at high temperature there is no indication that the axial and equatorial fluorines exchange their positions.

The second method is to make the central atom part of a ring system, that is, to use a chelate ligand. This not only introduces other means of studying intramolecular change, it allows them to be used on much more slowly reacting systems. The demands of ring strain (in addition to those of electronegativity) will now decide the relative stabilities of the possible arrangements. Thus,

$$\begin{array}{c} F \\ | \\ F-P \\ | \\ F \end{array} \begin{array}{c} CH_2-CH_2 \\ \diagdown \\ CH_2 \\ \diagup \\ CH_2-CH_2 \end{array}$$

behaves like $(CH_3)_2PF_3$ and does not show any indication of axial-equatorial fluorine exchange. On the other hand,

$$\begin{array}{c} F \\ | \\ F-P \\ | \\ F \end{array} \begin{array}{c} CH_2-CH_2 \\ \diagdown \\ | \\ \diagup \\ CH_2-CH_2 \end{array}$$

is only non-fluxional at low temperatures. It is thought that the strain in the five-membered ring raises the energy of the ground state so that it is not so different from the

$$\begin{array}{c} F \\ \diagdown | \\ P-CH_2-CH_2 \\ \diagup | \\ F \\ F \end{array} \begin{array}{c} CH_2-CH_2 \\ \diagdown \\ \end{array} ,$$

form that must be traversed if the axial and equatorial fluorines are to exchange positions. In the cyclic oxyphosphoranes, this situation can be reversed since the electronegative oxygens prefer to occupy the axial positions. The arrangement whereby the five-membered ring spans axial and equatorial positions is now stable and any isomerization requiring as intermediate a trigonal bipyramid in which both chelate donors are in the trigonal plane will require an increase in bond angle from 90° to 120° within the ring at phosphorus.

So far we have only mentioned the general features of the Berry twist but it may be of some value and even of some interest to consider one problem (and one problem only) in depth in order to gain some idea of the proper application. There are 20 ways of arranging five different ligands in a trigonal bipyramid and they comprise 10 enantiomeric pairs. Each pair can be represented in shorthand by listing the axial ligands and differentiated according to the order of arranging the remaining three equatorial ligands:

$$
\begin{array}{cccc}
\underset{\underset{15}{5}}{\overset{\underset{|}{1}}{\underset{3}{\nearrow}\text{M}-2,}} &
\underset{\underset{\overline{15}}{5}}{\overset{\underset{|}{1}}{\underset{4}{\nearrow}\text{M}-2,}} &
\underset{\underset{14}{4}}{\overset{\underset{|}{1}}{\underset{3}{\nearrow}\text{M}-2,}} &
\underset{\underset{\overline{14}}{4}}{\overset{\underset{|}{1}}{\underset{5}{\nearrow}\text{M}-2, \text{ and so on}}}
\end{array}
$$

A Berry twist is uniquely characterized by the pivot ligand and so the conversion

$$
\underset{5}{\overset{1}{\underset{3}{\nearrow}\text{M}-2}} \longrightarrow \underset{5\ 3}{\overset{1\ 4}{\text{M}-2}}
$$

can be represented by $15 \xrightarrow{2} 34$. Since there are three possible pivots, each isomer can be converted into three others as a result of such twists. The complete system can therefore be represented topologically by a regular closed 20 vertex three-dimensional figure in which each vertex (representing an isomer) is linked to the three others with which it is related by a Berry twist. The problem can be simplified by linking adjacent sites, say 1 and 2, by a chelate ligand. This will eliminate isomers $\underline{12}$ and $\overline{12}$ and leave 18 isomers and a system that can be represented by the 18 vertex polygon in Fig. 8-6.

The phosphorane derived from dimethyl phenylphosphonite and benzylidene-acetyl acetone (Fig. 8-7) is a useful example of this type of compound. The isomers designated as 15, $\overline{15}$, 13, and $\overline{13}$ are the only ones that satisfy both the electronegativity requirement (the axial positions should be filled by the most electronegative ligands) and the ring strain requirement (the five-membered ring should have a 90° angle at phosphorus). Because of the presence of an asymmetric carbon atom in the chelate the pairs are diastereoisomers rather than enantiomers and can be distinguished as such. 15 and $\overline{13}$ (also $\overline{15}$ and 13) are chemically identical but the labelling of the methoxy groups becomes important when we consider the n.m.r. Of the other possible isomers 35 and $\overline{35}$ maintain axial oxygens but contravene the ring strain requirement; 14, $\overline{14}$, 23, $\overline{23}$, 25, $\overline{25}$ satisfy the ring strain requirement but contravene the electronegativity

118 Stereochemical change   Ch. 8

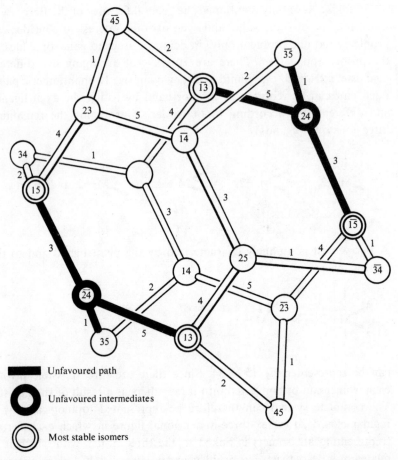

Fig. 8–6 The 18 possible isomers of [M (1, 2) 3, 4, 5] occupy the vertices of this regular figure. Connecting lines represent interconversion paths by the Berry twist. The path is labelled according to the pivot ligand

requirement; 34, $\overline{34}$, 45, and $\overline{45}$ contravene both requirements; and 24 and $\overline{24}$ have no axial oxygens.

Examination of Fig. 8-6 shows that none of the stable forms are adjacent and therefore directly interconvertible. Any change must go by way of one of the less stable isomers and the activation energy cannot be less than the energy difference between the stable forms and the least stable isomer along the pathway.

At this point then we can examine the facts. Two distinct changes are observed in the n.m.r. of the considered phosphorane when the temperature is raised. At 0°C the two methoxy groups become equivalent. This corresponds to the change $15 \rightleftarrows \overline{13}$ and $\overline{15} \rightleftarrows 13$, but at this temperature 15 and $\overline{15}$ and 13 and $\overline{13}$ are still quite distinct. It will be seen

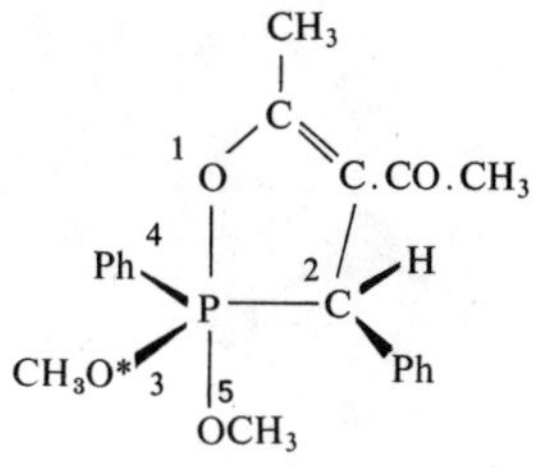

Fig. 8 – 7 Phosphorane derived from dimethyl phenylphosphonite and benzylideneacetylacetone. Showing labelling scheme and stable isomers

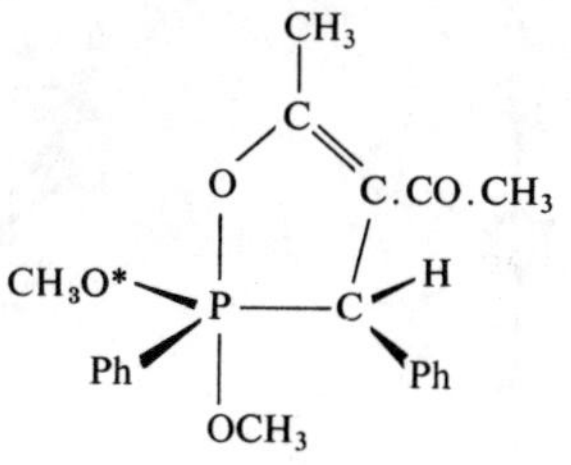

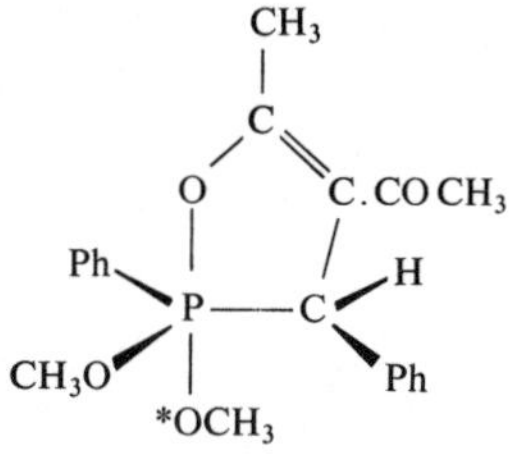

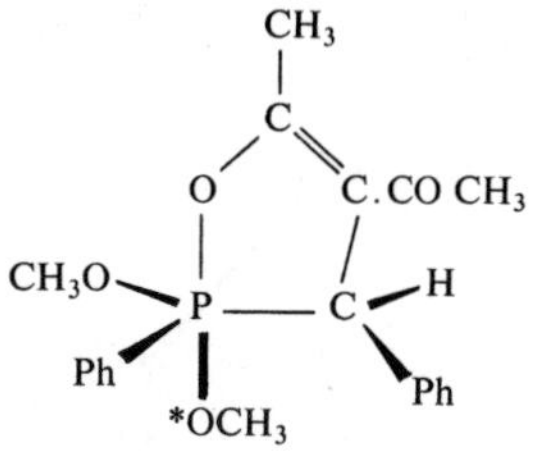

in Fig. 8-7 that in 15 and $\overline{13}$ the two phenyl groups are *trans*, whereas in $\overline{15}$ and 13 they are *cis*. The two signals for the proton in the asymmetric carbon merge at about 50°. This indicates that a path by way of 24 and $\overline{24}$, which would lead to simultaneous CH_3O and C—H broadening is not used at 0°. It is possible, but not necessary, that the barriers due to 24 and $\overline{24}$ are crossed at the 50° isomerization. The activation energies of these processes are in the region of 20 kcal.mol. and approach those related to bond breaking. However, in this case at least, ring opening can be observed as a new phenomenon when the temperature is raised still further.

This example has been chosen not because it is of special interest in its own right but because it presents a typical treatment of the approaches that have been adopted and the complexities that can be encountered in these problems of pseudorotation.

8-2-4 Pseudorotation in six-coordinate systems

In direct contrast to the observed behaviour of five-coordinate systems the octahedron is stereochemically extremely rigid. The most likely complimentary geometry for a pseudorotation is the trigonal prism and a trigonal twist of 60° about one of the threefold axes will convert an octahedron into a trigonal prism and vice versa (Fig. 8-8).

Apart from special cases, the trigonal prismatic arrangement is very rare indeed, although it can be forced where the steric requirements

120 Stereochemical change　　　　　　　　　　　　　　　　　　　　　　Ch. 8

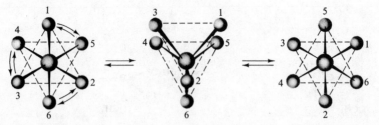

Fig. 8 – 8 Trigonal twist about a threefold axis of the octahedron

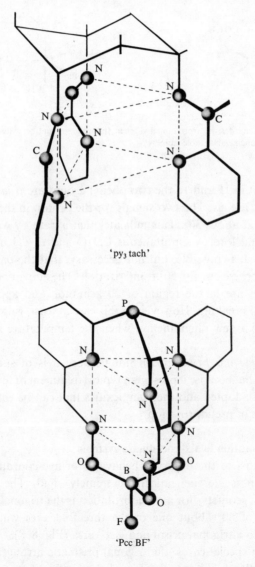

Fig. 8 – 9 Ligands that 'force' a trigonal prismatic geometry

of a multidentate ligand can only be satisfied in this way. (In general, a compromise is achieved somewhere between the two regular geometries.) Two interesting examples are shown in Fig. 8-9. In the case of PccBF, the tetrahedral boron that locks the geometry is put on after the metal has been coordinated. The presence of six sulphur donors sometimes can tip the balance completely, as in the case of MoS_2 (which is perhaps cheating) and $Re(S_2C_2Ph_2)_3$, and in many other cases there is partial distortion. It is probable that pseudorotation, by way of a trigonal twist, will be quite facile in this type of complex. However, with the types of octahedral complexes one normally encounters, the barrier to this type of pseudorotation is generally similar to or greater than the activation energy for bond fission. Examples of intramolecular isomerization in which a chelate ring opens and reconnects with steric change have been discussed earlier in this chapter. A great deal has been written about the various pseudo-rotations, or 'twists' for tris-chelated octahedral species and their relative merits have been considered. Two of the most common, as applied to $M(AB)_3$, are shown in Fig. 8-10(a) and (b). The distinction between twist and ring opening mechanisms is not always an easy one to make, especially when more than one path is available for the reaction.

8-2-5 Coordination numbers greater than six

In seven, eight, and nine-coordinate systems there are a number of regular geometries of fairly similar energies. Secondary features, such as inter-

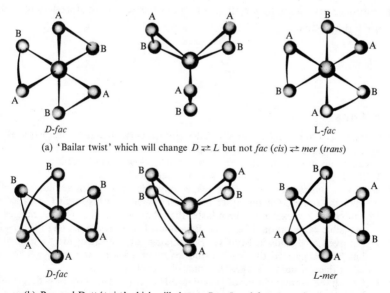

D-fac L-fac

(a) 'Bailar twist' which will change $D \rightleftarrows L$ but not *fac* (*cis*) $\rightleftarrows$ *mer* (*trans*)

D-fac L-mer

(b) Ray and Dutt 'twist' which will change $D \rightleftarrows L$ and *fac* $\rightleftarrows$ *mer* simultaneously

Fig. 8 – 10

ligand repulsion, start to assume a dominant role. Under these circumstances the stereochemistry is expected to be non-rigid and relatively easy pseudorotation, as was encountered in the five-coordinate systems, will be likely. The amount of detailed work on such systems is small at the moment.

8-2-6 Special case of hydride complexes

It has already been suggested that the 'pseudorotation' of NH_3 is better looked upon as a quantum mechanical 'tunnelling' by hydride and it is found that very many hydride complexes are stereochemically non-rigid. Thus $H Co[PPh(OEt)_2]_4$ has a distorted tetrahedral arrangement of the four phosphorus atoms about the cobalt with the hydrogen on one of the threefold axes above a face, which has opened out slightly to receive it (alternatively described as a trigonal bipyramid with H axial and with Co displaced some 0·492Å out of the trigonal plane away from the hydrogen). However, the signal from the hydride proton is split into a quintet ($^1H-^{31}P$ spin-spin coupling with four equivalent phosphorus atoms) and it is thought that the hydrogen can tunnel from face to face. A wide range of analogous compounds behave in a similar way. Octahedral cis-MH_2L_4 (M = Fe(II), Ru(II); L = PR_3) have temperature dependent n.m.r. spectra and can be shown to undergo rearrangement (all four phosphorus atoms appear equivalent) without any metal–ligand bond fission and there is some discussion as to the extent to which the hydrogens actually occupy specific coordination sites. H_4OsL_3, H_5ReL_3, ReH_9^{2-} (L = phosphines) all have n.m.r. spectra consistent with stereochemically non-rigid systems.

Problems

8-1 The ^{19}F n.m.r. spectrum of $(CH_3)_2PF_3$ indicates two magnetically inequivalent types of fluorine even at high temperatures whereas Cl_2PF_3 and Br_2PF_3 only show this non-equivalence at temperatures below $-60°C$. Explain.

8-2 Studies of the exchange of oxygen between $Co(C_2O_4)_3^{3-}$ and $H_2^{18}O$ labelled solvent indicate that six of the twelve oxygens exchange much more rapidly than the others. On the other hand, all twelve oxygens in $Cr(C_2O_4)_3^{3-}$ are equivalent and exchange much faster than the complex decomposes. Suggest a plausible explanation. [See J. A. Broomhead, J. Chem. Soc., (A) 1971, 645.]

8-3 All five CO ligands in $Mn(CO)_5Br$ exchange at equivalent rates although it has been suggested that carbon monoxide exerts a greater *trans* weakening effect than bromine. Suggest *two* plausible explanations.

8-4 Sketch the isomeric forms of $[Co\ dien(NH_3)_2Cl]^{2+}$ and show how they might be interconverted by trigonal twists. What restriction is placed upon the system by the tetrahedral secondary nitrogen? What would be the effect of replacing the ammonia ligands by ethylenediamine?

Bibliography

Cotton, F. A., Fluxional organometallic molecules. *Acc. Chem. Res.*, 1968, **1**, 257.

Mislow, K., The role of pseudorotation in the stereochemistry of nucleophilic displacement reactions. *Acc. Chem. Res.*, 1970, **3**, 321.

Muetterties, E. L., Stereochemically non-rigid structures. *Acc. Chem. Res.*, 1970, **3**, 266.

Ugi, I., D. Marquarding, H. Klusacek, P. Gillespie, and F. Ramirez. *Acc. Chem. Res.*, 1971, **4**, 288.

9 Oxidation and reduction

9-1 Introduction

Until now we have been concerned with changes in the composition and geometry of the coordination shell of the complex. Provided we keep away from areas where the question 'When is a bond not a bond?' (or vice versa) is asked, the description of a coordination shell and its geometry is readily validated in a crystalline solid by X-ray diffraction and less readily, but still convincingly, validated in solution by a wide range of the 'more sporting' techniques. Therefore we can discuss substitution and stereochemical change without losing contact with reality. The concept of oxidation state (and oxidation number) on the other hand is far less firmly based upon reality and depends very much upon the model that one uses to account for the bonding. This should not detract from the usefulness of the concept when it is applied to the classification and rationalization of chemical compounds and this usefulness should remain for many years to come. If one wishes to define oxidation-reduction reactions in terms of changes in oxidation state one should do so with the above reservations in mind. A very large number of reactions can be found in which there is a direct transfer of electrons, and only electrons, from the reductant to the oxidant. These reactions present no problems of classification whatsoever, especially when the electrons come from and go to essentially non-bonding orbitals. It is when atoms or groups of atoms are transferred that the problems arise. The reaction $SO_3^{2-} + ClO^- \rightarrow SO_4^{2-} + Cl^-$ has already been mentioned in this context. Is this a two-electron transfer from sulphur to chlorine or is it a nucleophilic attack by the sulphite group upon the oxygen of the hypochlorite? It is, of course, equally both. It represents a very wide group of reactions where it is not so much a question of electron transfer from reductant to oxidant but rather a change in the role of a pair of electrons from non-bonding to bonding, or vice versa. In this way we can understand how iridium(I) in *trans*-[Ir CO Cl(PPh$_3$)$_2$] can be oxidized to iridium(III), as [IrCO H$_2$Cl(PPh$_3$)$_2$] by H$_2$, which one normally thinks of as having reducing properties (except when it meets a better reducing agent, for example, $2Na + H_2 \rightarrow 2NaH$) and how the oxidation of Co(I) to Co(III) by addition of methyl iodide is best looked upon as a displacement of the iodide from the carbon by the Co(I) acting as a nucleophile. If two of the original eight non-bonding electrons are used for this purpose the cobalt changes formally from d^8 Co(I) to d^6 Co(III).

These aspects of oxidation-reduction detract from the simple principles and will not be considered further at this stage. They are of sufficient importance however to warrant separate consideration in their own right.

9-2 Electron transfer

In general, free solvated electrons in a condensed phase represent highly reactive chemical systems, and, if a suitable substrate is not present, the solvent itself will be reduced. Occasionally the solvent is inert to reduction (for example, liquid ammonia) and the properties of reasonably concentrated solutions of electrons can be studied at leisure but, as a rule, it is necessary to generate the solvated electron and then study its disappearance, all within a very small fraction of a second. The techniques for doing this exist and a great deal is now known about the kinetics of reduction by the solvated electron, nevertheless oxidation and reduction reactions do not normally proceed by way of the formation and the consumption of the solvated electron and direct bimolecular interaction between the oxidant and the reductant is involved in all of the cases that have been studied. Because of the synchronous nature of the process the term **redox** has been applied to these reactions.

It is possible to conceive of two types of redox process. In the first, commonly termed **Outer Sphere** reactions, the interaction between the oxidant and the reductant at the time of electron transfer is small and they go through the process with their coordination shells intact. The second type of reaction, which is termed **Inner Sphere**, requires that the oxidant and reductant are firmly bonded during the act of electron transfer and that a bridge formed by at least one ligand that is common to the coordination of shells of the oxidant and the reductant serves as the channel through which the electron is transmitted. In this type of mechanism there is, of necessity, a sequence of steps and it is not necessarily the actual redox process that is rate determining.

9-3 Reactions of the solvated electron

If a pulse of high energy electrons (5–15 MeV) from a linear accelerator is passed through water, the reaction

$$H_2O \xrightarrow{\sim\!\sim\!\sim} HO\cdot + H^+ + e^-_{aq}$$

takes place and it is possible to achieve stationary millimolar concentrations of the solvated electron. In the absence of reducible substrates this solvated electron will react with water

$$e^-_{aq} + H_2O \longrightarrow H\cdot + OH^- \quad k = 16 \, M^{-1} \, s^{-1} \quad (20\text{–}25°)$$
$$e^-_{aq} + H_3O^+ \longrightarrow H\cdot + H_2O \quad k = 2\cdot 3 \times 10^{10} \, M^{-1} \, s^{-1} \quad (20\text{–}25°)$$

Table 9-1 Second-order rate constants for reduction of metal complexes by e^-_{aq} ($10^{-9} k_2$ at 25° $M^{-1} s^{-1}$)

Complex	$10^{-9}k_2$	Complex	$10^{-9}k_2$	Complex	$10^{-9}k_2$
$Al(H_2O)_6^{3+}$	2·0	$Al\ EDTA^-$	0·03		
$Cr(H_2O)_6^{3+}$	60	$Cr\ EDTA^-$	26	$Cr(CN)_6^{3-}$	15
$Cr(H_2O)_6^{2+}$	42				
$Mn(H_2O)_6^{2+}$	0·08	$Mn\ EDTA^{2-}$	<0·002	$Mn(CN)_6^{4-}$	5
$Fe(H_2O)_6^{2+}$	0·12			$Fe(CN)_6^{4-}$	<0·0001
$Co(NH_3)_6^{3+}$	90				
$Co(H_2O)_6^{2+}$	12	$Co\ EDTA^{2-}$	0·51	$Co(NO_2)_6^{3-}$	58
$Ni(H_2O)_6^{2+}$	22				
$Cu(H_2O)_6^{2+}$	33				
$Zn(H_2O)_6^{2+}$	15				
$Pr(H_2O)_n^{3+}$	0·29				
$Nd(H_2O)_n^{3+}$	0·59				
$Sm(H_2O)_n^{3+}$	25				
$Eu(H_2O)_n^{3+}$	61				
$Gd(H_2O)_n^{3+}$	0·55				
$Tm(H_2O)_n^{3+}$	3·0				
$Yb(H_2O)_n^{3+}$	43				

Diffusion control for M^{3+} = 72   Diffusion control for M^{3-} = 1.6

but provided the hydrogen ion concentration is low enough, it is possible to examine the solvated electron with suitably rapid techniques. Thus, the spectrum has been measured and shown to be very similar to that of dilute solutions of alkali metals in liquid ammonia and other similar systems, where the reaction between the electron and the solvent can be very slow. By measuring the rate of disappearance of the solvated electron generated in the presence of suitably reducible substrates it is possible to study the kinetics of their reduction and, if the reduction leads to otherwise unknown or unstable products, to study the kinetics of the substitution or redox properties of these species. It is generally useful to generate the solvated electron in the presence of an hydroxyl radical trap such as methanol in order to prevent a parallel oxidation by $\cdot OH$. (Alternatively, the solvated electron may be trapped, by N_2O, and the oxidizing properties of the $\cdot OH$ radical used to generate an unstable high oxidation state. Thus, the reactions of $Cu^{3+}aq$, generated by the oxidation $Cu^{2+}aq + \cdot OH \longrightarrow Cu^{3+}aq + OH^-$, can be followed by pulse radiolyses.) A number of second-order rate constants and activation energies for the reduction of the same metal aquo ions and other complexes are collected in Table 9-1 where it will be seen that (a) many, but by no means all, of these reductions are fast, the rate approaching that of a diffusion controlled process, and (b) there often seems to be a parallel between the rate of reaction and the redox potential associated with the reduction. This is qualitatively true in the case of the lanthanide ions where the tervalent ions of the elements, Sm, Eu, Tm, Yb, which are known to form divalent compounds, Eu and Yb especially, reduce at rates approaching diffusion control while the others are reduced much more slowly. It is also claimed that the rates of reduction of the bivalent ions of the first row transition series parallel the trends in the second ionization potential of the gaseous ions.

Arguments based on energetics are somewhat dangerous in this context because, where measured, the activation energies are all in the same region (~ 3.5 kcal.mol^{-1}). This is true even for the slower reactions and it has been suggested that the activation energy is a property of the hydrated electron alone and that it is something it requires to prepare itself for transfer. The observed activation energy is also typical for a diffusion controlled process in water. Since the reducing power of the hydrated electron has been estimated as being equivalent to a redox potential of about 2·7 volts most of the reductions will be strongly exothermic and so the Franck-Condon restrictions (see below) will be unimportant.

9-4 Outer sphere redox reactions

The process whereby one or more electrons are transferred between two encountering molecules (whether in the gas phase or in solution) has been a subject of considerable theoretical as well as practical interest. In order to establish this mechanism it was necessary to find examples where the act of redox was very much faster than any substitution process that the oxidant or reductant would undergo and many such systems exist. Thus the reaction, $[Fe(Me_2bipy)_3]^{2+} + [Fe(phen)_3]^{3+} \rightarrow [Fe(Me_2bipy)_3]^{3+} + [Fe(phen)_3]^{2+}$ (Me_2bipy = 4,4′-dimethyl-2,2′-bipyridine and phen = 1,10-phenanthroline) has a second-order rate constant greater than 10^8 M^{-1} s^{-1} at 25°, which means that if 10^{-4} M solutions of the reagent are mixed, >99% of the redox reaction is complete in 13 ms. On the other hand, the half-life for substitution in $Fe(phen)_3^{3+}$ is $1\cdot4 \times 10^4$ s at 25° and the lability of $Fe(Me_2bipy)_3^{2+}$ is probably not much different. One could also see no transfer of ligand between oxidant and reductant and presumably an experiment could be carried out in which the reaction took place in the presence of ^{14}C-labelled free ligand in order to show that none was incorporated in the course of the redox process. Although the unequivocal demonstration of coordination shell retention requires a set of reagents which undergo redox reactions much faster than they undergo substitution, it must not be assumed that this is a property of the outer sphere redox mechanism. Very many cases exist where the reagents are substitutionally labile and yet the redox process is outer sphere. The problem of distinguishing between inner and outer sphere mechanism when both oxidant and reductant and the products are substitutionally labile will be considered later in the chapter. The simplest case to consider is that in which there is no net chemical change, for example, a reaction between two complexes in which the same central atoms have identical coordination shells but are in different oxidation states, viz.

$$[\overset{*}{Mn}^{VII}O_4]^- + [Mn^{VI}O_4]^{2-} \rightleftharpoons [\overset{*}{Mn}^{VI}O_4]^{2-} + [Mn^{VII}O_4]^-$$

$$[\overset{*}{Fe}^{III}(CN)_6]^{3-} + [Fe^{II}(CN)_6]^{4-} \rightleftharpoons$$
$$[\overset{*}{Fe}^{II}(CN)_6]^{4-} + [Fe^{III}(CN)_6]^{3-}$$

$$[\overset{*}{Os}^{II}(bipy)_3]^{2+} + [Os^{III}(bipy)_3]^{3+} \rightleftharpoons$$
$$[\overset{*}{Os}^{III}(bipy)_3]^{3+} + [Os^{II}(bipy)_3]^{2+}$$

In such reactions, the act of electron transfer does not change the concentration of any of the species and therefore involves no change in free energy (that is, ΔG^0 for such a process is zero if we disregard any effect arising from the means adopted to measure the rate such as the introduction of a small amount of isotopically labelled species). Indeed, the question arises as to how one can measure the rates of such processes. The obvious way, and one used most commonly in the early days of the interest in this type of reaction, was to add a small amount of one of the reagents labelled by a radioactive isotope of the central atom and then separate the two forms to see how rapidly the isotopic label was distributed between the two forms. Thus, if radioactive ^{54}Mn labelled $KMnO_4^-$ is added to a mixture of $KMnO_4$ and $K_2MnO_4^-$ and then (rapidly in this case) the MnO_4^- is precipitated selectively as $Ph_4As^+MnO_4^-$ or the MnO_4^{2-} as $BaMnO_4$, the distribution of radioactivity can easily be measured. In the late fifties, when radio isotopes and the techniques for their use became readily available, this type of study became very popular indeed. In special cases, other techniques may be used. Thus ^{17}O n.m.r. studies of $MnO_4^- - MnO_4^{2-}$ mixtures as a function of temperature and concentration can provide the necessary kinetic data for the electron transfer process. A most elegant method was developed by Dwyer for tris-chelated complexes that could be resolved and remain optically stable. The reaction

$$D\text{-}[Os(bipy)_3]^{2+} + L\text{-}[Os(bipy)_3]^{3+} \rightleftharpoons$$
$$D\text{-}[Os(bipy)_3]^{3+} + L[Os(bipy)_3]^{2+}$$

will lead, at equilibrium, to the D form and the L form being distributed equally between the two oxidation states. The act of electron transfer will lead to a change in the total optical rotation and, if equal concentrations of reagents are used, the rate of racemization will be a measure of the rate of electron transfer.

9-5 Franck–Condon restrictions

The electron transfer processes must satisfy the Franck–Condon restrictions which arise, in principle, from the fact that the act of electron transfer takes place within a time ($\sim 10^{-15}$ s) that is very much shorter than the time required for nuclei to change their positions ($> 10^{-13}$ s). This means that the nuclei remain frozen in position during the act of electron transfer and two important consequences follow. The first is that no angular momentum can be transferred to or from the transition state during the act of electron transfer and there will be restrictions on changes in spin angular momentum. Thus, the reaction

$$[Co(phen)_3]^{2+} + [Co(phen)_3]^{3+} \rightleftharpoons [Co(phen)_3]^{3+} + [Co(phen)_3]^{2+}$$
$$t_{2g}^6 e_g^1 \qquad\quad t_{2g}^6 \qquad\qquad t_{2g}^6 \qquad\quad t_{2g}^6 e_g^1$$

which has a rate constant of $1 \cdot 1 M^{-1} s^{-1}$ at $25°$ simply requires the transfer of electron from the e_g orbital of one cobalt to that of the other. On the other hand, the reaction

$$Co(NH_3)_n^{2+} + Co(NH_3)_6^{3+} \rightleftharpoons Co(NH_3)_n^{3+} + Co(NH_3)_6^{2+}$$
$$t_{2g}^5 e_g^2 \qquad\quad t_{2g}^6 \qquad\qquad t_{2g}^6 \qquad\quad t_{2g}^5 e_g^2$$

which is very slow, k = 10^{-9} M^{-1} s^{-1} at 25°, requires a change of spin multiplicity as well. The restriction is not absolute and can be greatly reduced by a small amount of spin-orbit coupling.

The second consequence is that the oxidant and the reductant must reorganize themselves before the act of electron transfer in a way that ensures that the energies of the oxidant and reductant in their transition states are identical. This can be clearly understood if we consider the reaction

$$\text{Fe}(\text{H}_2\text{O})_6^{2+} + \text{Fe}(\text{H}_2\text{O})_6^{3+} \rightleftharpoons \text{Fe}(\text{H}_2\text{O})_6^{3+} + \text{Fe}(\text{H}_2\text{O})_6^{2+}$$

Let the energy dependence on bond length be represented by the two-dimensional potential energy *vs* Fe—O distance diagram in Fig. 9-1. The ground state Fe—O bond length in Fe(II) (2·21 Å) is longer than that in Fe(III) (2·05 Å) and so if electron transfer took place from Fe(H$_2$O)$_6^{2+}$ in its ground state to Fe(H$_2$O)$_6^{3+}$ in its ground state the product would be a compressed Fe(II) ion and a stretched Fe(III) ion. These would both be vibrationally excited and release their energy to their surroundings. If this were allowed, a solution containing ferrous and ferric ions would get warmer and, while this would be a welcome phenomenon that would rapidly solve all the world's problems of dwindling sources of energy, it unfortunately contravenes umpteen laws of thermodynamics. Consequently, the reagents must match their energies by bond compression, stretching, and bending before electron transfer can occur. It has been calculated that in order to equalize the bond lengths at 2·09 Å, an enthalpy

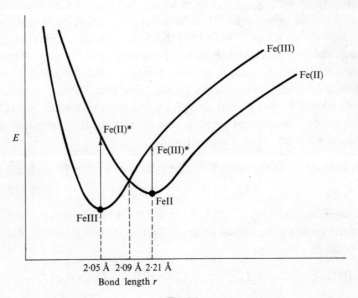

Fig. 9-1

of activation of 5·6 kcal.mol^{-1} would be required. The observed value of 10·5 kcal.mol^{-1} indicates that this is not the only restriction.

In addition to the contributions from bond energy and spin multiplicity matching, the activation energy will also contain terms resulting from the coming together of charged species and the necessary rearrangement of the solvation shells. At this point the different theoretical approaches diverge. In one, the electron transfer is looked upon in terms of tunnelling through a barrier whereby the electron changes from one energy surface to another. The transmission coefficient for tunnelling, κ, will appear in the equation for the rate constant

$$k = \kappa \frac{RT}{Nh} e^{-\Delta G^{\ddagger}/RT}$$

(where N = Avogadro's number and h = Planck's constant) and will generally increase as the reagents get closer together. However, the closer they approach the greater the repulsion and hence $\Delta G^{\ddagger}$ and so an optimum separation must be achieved to maximize k.

An alternative approach considers what is termed an 'adiabatic' electron transfer in which a single potential energy surface is used to describe the transferring electron and the interaction between orbitals in the oxidant and the reductant, while small (that is, little formal bonding) is sufficient to lower the energy barrier to allow the electron to move freely through it. Again, the adjustment of bond energies, solvation shells, and so on are necessary contributions to the free energy of activation.

Because of the sensitivity of rate to the energetics of approach of the reagents, outer sphere reactions are very sensitive to the nature of other species in solution. Thus, the $MnO_4^- - MnO_4^{2-}$ reaction is sensitive to the nature of the cations that must, of necessity, be present. For the alkali metal ions, $Cs^+ > K^+ > Na^+ > Li^+$ in terms of catalytic effect. At 0°C in 0·16M MOH the rate constant, when M = Cs (2·47 × $10^3 M^{-1} s^{-1}$), is more than three times greater than the value when M = Na (7·3 × $10^2 M^{-1} s^{-1}$). Similarly the $Fe(CN)_6^{4-} - Fe(CN)_6^{3-}$ reaction is extremely sensitive to the nature of the cations present. Anions likewise can catalyse the reaction between cationic species even when they do not enter the coordination shell. In order then to obtain data that can serve to check the theoretically based calculations it is essential to ensure that this type of catalysis is avoided or that the calculation takes account of the effect.

When the act of electron transfer leads to a net change in the free energy of the system the consequences of the Franck-Condon restriction can become less severe. For reactions that are exothermic the requirement that the energies of the oxidant and reductant must be completely matched in the transition state can be relaxed because some of the

Table 9-2 Second-order rate constants for some outer sphere redox reactions (25°, water)

Reagents	Electron configuration	$k_2 M^{-1} s^{-1}$
(a) Zero free energy change		
$Fe(phen)_3^{2+} + Fe(phen)_3^{3+}$	$t_{2g}^6 + t_{2g}^5$	10^5
$Os(bipy)_3^{2+} + Os(bipy)_3^{3+}$	$t_{2g}^6 + t_{2g}^5$	5×10^4
$Fe(CN)_6^{4-} + Fe(CN)_6^{3-}$	$t_{2g}^6 + t_{2g}^5$	7.4×10^2
$IrCl_6^{3-} + IrCl_6^{2-}$	$t_{2g}^6 + t_{2g}^5$	10^3
$Co(phen)_3^{2+} + Co(phen)_3^{3+}$	$t_{2g}^6 e_g + t_{2g}^6$	1.1
(b) Gain in free energy		
$Fe(CN)_6^{4-} + Fe(phen)_3^{3+}$	$t_{2g}^6 + t_{2g}^5$	10^8
$Fe(CN)_6^{4-} + IrCl_6^{2-}$	$t_{2g}^6 + t_{2g}^5$	3.8×10^5
$Ru(phen)_3^{2+} + RuCl_6^{2-}$	$t_{2g}^6 + t_{2g}^5$	2.5×10^9

redistribution energy can be accepted in the heat of the reaction. Such reactions tend to be considerably faster. On the other hand, an endothermic process will suffer greater restrictions. A collection of some second-order rate constants for some proven outer sphere redox reactions is shown in Table 9-2. One property of the complexes shown in this table is the potential ability of the ligands to 'conduct' electrons between the non-bonding orbitals of the metal and the periphery of the complex. Even in the case of $IrCl_6^{2-}$, e.s.r. measurements indicate clearly the presence of part of an unpaired electron on the chlorine ligands, and it has been suggested that electron transfer from the ligand to the empty t_{2g} orbital on the t_{2g}^5 Ir(IV) could account for this. It should not be assumed that electron conductance in the ligand is a requirement for an outer sphere mechanism. Reagents with non-conducting ligands such as NH_3 or H_2O can participate in outer sphere reactions but, in general, they do so at a very much slower rate.

In general, the rates of outer sphere electron transfers involving a net standard free energy change will relate quite closely to the magnitude of this change. An approximation by Marcus allows the rate constant, k_{12}, for the process

$$Ox_{(1)} + Red_{(2)} \xrightarrow{k_{12}} Red_{(1)} + Ox_{(2)}$$

to be obtained from the equilibrium constant of the above reaction, K_{12}, and the rate constant for the symmetrical reactions

$$Ox_{(1)} + Red_{(1)} \rightleftharpoons Red_{(1)} + Ox_{(1)}, \; k_1$$

and

$$Ox_{(2)} + Red_{(2)} \rightleftharpoons Red_{(2)} + Ox_{(2)}, k_2$$

using the expression

$$k_{12} = (k_1 k_2 K_{12} f)^{1/2}$$

f is generally close to unity and given by the expression

$$\log f = (\log K_{12})^2 / 4 \log(k_1 k_2 / Z^2)$$

(Z is a measure of the number of collisions per unit time and volume). This works reasonably well in many cases and quite often a significant discrepancy between the calculated and experimental rate constants indicates that an alternative mechanism (usually inner sphere) has taken over.

9-6 Inner sphere redox reactions

In the outer sphere mechanism just described, the oxidant and reductant have no real communication one with the other to help them equalize their energies as required by the Franck–Condon restriction. There may be a small amount of orbital overlap to give some guidance but if there were an atom, or group of atoms, firmly bonded both to the oxidant and the reductant this would provide not only a means of correlating energies but also a path along which the electrons may travel.

The demonstration of such a mechanism in which a firm ligand bridge was shown to be present at the time of electron transfer was made most elegantly by Taube. He chose a reacting system whereby the reduced forms of the redox system were substitutionally labile and the oxidized forms substitutionally inert. In this way the bridging ligand passed from the oxidant to the oxidized form of the reductant.

The classical reaction was

$$Co(NH_3)_5 Cl^{2+} + Cr(H_2O)_6^{2+} \xrightarrow{H^+} Co_{aq}^{2+} + 5NH_4^+ + ClCr(H_2O)_5^{2+}$$

Co(III) inert   Cr(II) labile   Co(II) labile   Cr(III) inert

Careful isolation of the product showed that virtually all of the Cr(III) was in the form of $Cr(H_2O)_5 Cl^{2+}$ and, working with radioactive chloride, it could be shown that all of the chloride came from the original $Co(NH_3)_5 Cl^{2+}$ and none from free chloride ion in solution. Since the cobalt(III) complex could not lose its chloride before being reduced and since the chromium after it was oxidized could not pick it up, it was demonstrated quite unambiguously that the chloride was transferred in the act of oxidation–reduction and must therefore have been bonded to both Co and Cr at this time.

This method of demonstrating the existence of the bridge had led

134 Oxidation and reduction

at one time or another to a little bit of confusion between the inner sphere mechanism and processes involving monovalent atom transfer from oxidant to reductant (one-electron transfer in the opposite direction) and bivalent atom transfer (two-electron transfer in the opposite direction) which can also be depicted as nucleophilic substitution by the reductant (or electrophilic substitution by the oxidant). The transfer of the bridged ligand is not a requirement of the mechanism, it is purely a convenient way of demonstrating its existence.

The general mechanism can be formulated in terms of a sequence of steps

(1) $Red^{\bullet}_{(1)} + X-Ox_{(2)} \longrightarrow Red^{\bullet}_{(1)}-X-Ox_{(2)}$ bridge formation

or (1') $Red^{\bullet}_{(1)}-X + Ox_{(2)} \longrightarrow Red^{\bullet}_{(1)}-X-Ox_{(2)}$ (very uncommon)

(2) $Red^{\bullet}_{(1)}-X-Ox_{(2)} \longrightarrow Ox_{(1)}-X-{}^{\bullet}Red_{(2)}$ electron transfer

(3) $Ox_{(1)}-X-{}^{\bullet}Red_{(2)} \longrightarrow Ox_{(1)}-X + {}^{\bullet}Red_{(2)}$
    or $Ox_{(1)} + X-{}^{\bullet}Red_{(2)}$ } bridge breaking
    or $Ox_{(1)} + {}^{\bullet}Red_{(2)} + X$

Processes (1) and (3) are nothing more complicated than ligand substitution reactions and only (2) involves any change in oxidation state. Occasionally it is possible to find a two-stage electron transfer process with the ligand mediating:

$$Red^{\bullet}_{(1)}-X-Ox_{(2)} \longrightarrow Ox_{(1)}-\overset{\bullet}{X}-Ox_{(2)}$$

$$Ox_{(1)}-\overset{\bullet}{X}-Ox_{(2)} \longrightarrow Ox_{(1)}-X-\overset{\bullet}{Red}_{(2)}$$

The evidence for this type of process will be discussed separately.

In principle, any of these steps either simply or in combination can be rate-determining and it is possible, from the kinetics and pattern of behaviour, to say where the important barrier to reaction lies. It is convenient, and instructive, to consider the different possibilities.

9-6-1 Bridge formation is rate determining

If the substitutional lability of the substrates is less than the rate of electron transfer and if, for some reason, the outer sphere mechanism does not take over, then the rate of redox reaction may be controlled by the rate at which the bridging ligand can penetrate the coordination shell of the more labile component. As a rule, the process will be an anation and follow the usual simple second-order rate law that is found when the pre-equilibrium association is small. Thus the kinetic form alone will not distinguish rate-determining bridge formation from the other possibilities, but the rate constant will. This will be virtually independent of the nature of the less labile components and certainly independent of the overall free

Inner sphere redox reactions

Table 9-3 Second-order rate constants for the oxidation of V^{2+} aq in water at 25°

Oxidant	$k_2 M^{-1} s^{-1}$	$\Delta H^{\ddagger}$ kcal.mol^{-1}	$\Delta S^{\ddagger}$ cal.deg^{-1}.mol^{-1}
Cu aq^{2+}	26·6	11·4	−13·8
Co(NH$_3$)$_5$Cl^{2+}	7·6	7·4	−30
Co(NH$_3$)$_5$Br^{2+}	25	9·1	−22
Co(NH$_3$)$_5$N$_3^{2+}$	13	11·7	−14
Co(NH$_3$)$_5$C$_2$O$_4$H^{2+}	12·5	12·2	−13
cis-[Co en$_2$NH$_3$N$_3$]$^{2+}$	10·3	12·6	−12
cis-[Co en$_2$H$_2$ON$_3$]$^{2+}$	16·6	12·1	−12
Co(CN)$_5$N$_3^{3-}$	110		
Fe aq^{3+}	1·8 × 10^4		
FeCl aq^{2+}	4·6 × 10^5		
Ru(H$_2$O)$_5$Cl^{2+}	1·9 × 10^3		
Ru(NH$_3$)$_5$O$_2$CCH$_3^{2+}$	1·3 × 10^3		
Ru(NH$_3$)$_5$Cl^{2+}	3·0 × 10^3	3·0	−30
Ru(NH$_3$)$_5$Br^{2+}	5·1 × 10^3	2·8	−34
Ru(NH$_3$)$_5$py^{3+}	1·2 × 10^5		
IrCl$_6^{2-}$			

energy change for the total reaction. Examples of this can be found in the reactions involving V^{2+} aq (d^3, in which the rate constant for water exchange is 100 s^{-1} at 25°) and Ru(NH$_3$)$_5$H$_2$O^{2+} (d^6, $k_{exch} \sim 10$ s^{-1}) and the effect is most clearly shown in the reductions by V^{2+} aq (Table 9-3). Two groups of data are collected, the first of which contains the rate constants which do not depend greatly upon the nature of the oxidant and compare very closely with rate constants for anation in V^{2+} aq. These correspond to reactions in which the formation of the bridge between the oxidant and the reductant is the rate-determining step. The rate constants in the second group are generally much larger and an outer sphere mechanism is operating. It is interesting (but probably not rewarding) to speculate on the mechanisms of the reduction of the chloro- and bromo-pentammine cobalt(III) cation which, while appearing to be in an inner sphere mechanism group on the basis of their rate constants, seem not to fit if their activation parameters are considered.

9-6-2 Electron transfer within the bridged intermediate is rate-determining

When at least one component is sufficiently substitutionally labile the act of forming the bridge will be fast and reversible. Indeed, this may be looked upon as a rapidly attained pre-equilibrium followed by a relatively slow 'rearrangement':

$$\text{Red}^\bullet_{(1)} + \text{X—Ox}_{(2)} \xrightleftharpoons{K} \text{Red}^\bullet_{(1)}\text{—X—Ox}_{(2)} \quad \text{fast}$$

$$\text{Red}^\bullet_{(1)}\text{—X—Ox}_{(2)} \xrightarrow{k} \text{Ox}_{(1)}\text{—X—}^\bullet\text{Red}_{(2)} \quad \text{slow}$$

$$\downarrow \text{fast}$$

$$\text{products}$$

Provided the thermodynamic stability of the bridged species is small the kinetic form will be of the simple second-order form associated with redox reactions, but if a significant fraction is in the form of the bridged species there will be a departure from the second-order kinetics. This has been observed but only as a rare occurrence.

The characteristic feature of this type of activation for an inner sphere mechanism is that the rate is very sensitive to the natures of the oxidant, the reductant, and the bridge. Some examples are collected in Table 9-4.

Table 9-4 Rate constants for some inner-sphere redox reactions where bridge formation and breaking is not rate-determining

Reductant	Oxidant	Bridge = X	$k_2(25°)\text{M}^{-1}\text{s}^{-1}$
Cr^{2+}	$\text{Co(NH}_3)_5\text{X}^{n+}$	CH_3COO	1.8×10^{-1}
Cr^{2+}	$\text{Co(NH}_3)_5\text{X}^{n+}$	—NCS	1.9×10^1
Cr^{2+}	$\text{Co(NH}_3)_5\text{X}^{n+}$	—CN	3.6×10^1
Cr^{2+}	$\text{Co(NH}_3)_5\text{X}^{n+}$	—F	2.5×10^5
Cr^{2+}	$\text{Co(NH}_3)_5\text{X}^{n+}$	—N$_3$	3×10^5
Cr^{2+}	$\text{Co(NH}_3)_5\text{X}^{n+}$	—Cl	6×10^5
Cr^{2+}	$\text{Co(NH}_3)_5\text{X}^{n+}$	—Br	1.4×10^6
Cr^{2+}	$\text{Co(NH}_3)_5\text{X}^{n+}$	—I	3.4×10^6
Cr^{2+}	$\text{Co(NH}_3)_5\text{X}^{n+}$	—PO$_4$	4.8×10^9
Cr^{2+}	$\text{Cr(NH}_3)_5\text{X}^{2+}$	—F	2.7×10^{-4}
Cr^{2+}	$\text{Cr(NH}_3)_5\text{X}^{2+}$	—Cl	5.1×10^{-2}
Cr^{2+}	$\text{Cr(NH}_3)_5\text{X}^{2+}$	—Br	3.2×10^{-1}
Co(CN)_5^{3-}	$\text{Co(NH}_3)_5\text{X}^{n+}$	—PO$_4$	5.2×10^2
Co(CN)_5^{3-}	$\text{Co(NH}_3)_5\text{X}^{n+}$	—NCS	1.1×10^6
Co(CN)_5^{3-}	$\text{Co(NH}_3)_5\text{X}^{n+}$	—N$_3$	1.6×10^6
Co(CN)_5^{3-}	$\text{Co(NH}_3)_5\text{X}^{n+}$	—Cl	5×10^7
Co(CN)_5^{3-}	$\text{Co(NH}_3)_5\text{X}^{n+}$	—Br	2×10^9
Co(CN)_5^{3-}	$\text{Co(NH}_3)_5\text{X}^{n+}$	—I	too fast
Fe^{2+}	$\text{Co(NH}_3)_5\text{X}^{n+}$	—Br	7.3×10^{-4}
Fe^{2+}	$\text{Co(NH}_3)_5\text{X}^{n+}$	—Cl	1.4×10^{-3}
Fe^{2+}	$\text{Co(NH}_3)_5\text{X}^{n+}$	—NCS	3×10^{-3}
Fe^{2+}	$\text{Co(NH}_3)_5\text{X}^{n+}$	—F	6.6×10^{-3}
Fe^{2+}	$\text{Co(NH}_3)_5\text{X}^{n+}$	—N$_3$	8.7×10^{-3}
Fe^{2+}	$\text{Co(NH}_3)_5\text{X}^{n+}$	—SCN	1.2×10^{-1}

In most of the cases cited, evidence has been presented to justify the assignment of an inner sphere mechanism. A number of lessons can be learnt from these data. For example, it can be seen how the rate depends upon the nature of the bridge, partly due to the relative affinities of the reductant for the bridging ligand (the effect controlling K) and partly due to the effect of the bridge as a mediator for electron transfer. The sequence of reactivity of the halide as a bridge to some extent reflects the class a–class b character of the system. Thus for reduction by Fe^{2+} which is very much a class a metal, the reactivity sequence is $F > Cl > Br$; for Cr^{2+} the sequence is reversed although the whole rate range covered is not much more than one order of magnitude. With the strongly class b reductant $[Co(CN)_5]^{3-}$ the sequence is $F \ll Cl \ll Br \ll I$ and it has been shown that the reaction with $Co(NH_3)_5F^{2+}$ ($k_2 = 2 \times 10^3 M^{-1} s^{-1}$ at 25°) is almost entirely an outer sphere process. A similar effect is seen in examining the way in which the bridging ability of the thiocyanate ligand depends upon the nature of its bonding. Thiocyanate is an ambident ligand with a class a (hard) end and a class b (soft) end. When bonded to the oxidant by way of nitrogen (the class a donor) it presents sulphur as the bridging donor. In reactions with Cr^{2+} the sulphur is not greatly welcome and isothiocyanato pentammine cobalt(III) complex is a much less effective oxidant than the otherwise similar azido complex. However, with the class b $Co(CN)_5^{3-}$ reductant, the $Co(NH_3)_5NCS^{2+}$ oxidant is very effective. The comparison of the effectiveness —N bonded and S—bonded isomers in the oxidation of Fe^{2+} is interesting, the one presenting the nitrogen to the Fe^{2+} being some 40 times more effective. The electronic conductivity of a polyatomic bridge is of some importance since if there is a suitable conjugated path through the system, the ligand *may* allow remote attack, thereby reducing repulsive interaction between the oxidant and the reductant and this *may* result in a highly reactive system. The actual ability of the bridge to act as an electron acceptor also plays a part. The problem can be looked upon, in part, as one of tunnelling through the energy barrier making use of the bridge, and in part as transmission of the electron through the lowest empty molecular orbital of the bridge. These two models are not totally unrelated. The residence time of the electron on the bridge may become chemically significant.

A major feature of the process when bridge formation is not rate-determining is the sensitivity of rate to the overall free energy change in the reaction. Thus, Co(III) is a much better oxidizing agent than Cr(III) and this is reflected in the ratio of the rate constants for

$$Cr^{2+} + M(NH_3)_5X^{2+} \longrightarrow M^{2+} + Cr(H_2O)_5X^{2+} + 5NH_3$$

where the ratio $k_2(Co)/k_2(Cr) = 10^9$ ($X = F$), 10^7 ($X = Cl$), 5×10^6 ($X = Br$). Likewise Fe^{2+} is a poorer reducing agent than Cr^{2+}.

9-6-3 Slow bridge fission in the redox product

In the extreme the product of the reaction can be sufficiently inert to retain the bridge. Thus the reaction between $Co(CN)_5^{3-}$ and $Fe(CN)_6^{3-}$ gives the complex anion $[(CN)_5CoNCFe(CN)_5]^{6-}$ which can be obtained as a crystalline salt. Here the properties are completely consistent with the presence of a t_{2g}^6 Co(III) and a t_{2g}^6 Fe(II). Indeed, further oxidation forms the isolatable $[(CN)_5CoNCFe(CN)_5]^{5-}$ which contains Co(III) and Fe(III). More commonly, the bridge is reasonably labile and the bridging product can only be observed as a transient. Thus in the reaction of

$$[(NH_3)_5Ru-N\underset{}{\bigcirc}-\underset{NH_2}{\overset{O}{C}}]^{3+}$$

with Cr^{2+}, the intermediate

$$[(NH_3)_5Ru-N\underset{}{\bigcirc}-\underset{NH_2}{C}=O-Cr(H_2O)_5]^{5+}$$

can be seen slowly aquating to

$$[(NH_3)_5Ru-N\underset{}{\bigcirc}-CO(NH_2)]^{2+} + Cr(H_2O)_6^{3+}$$

This is one of the few cases where the bridge remains with the reduced form of the oxidant. If the overall free energy change is not too great, the act of electron transfer will be reversible:

$$\overset{\bullet}{Red}_{(1)} + X-Ox_{(2)} \rightleftharpoons \overset{\bullet}{Red}_{(1)}-X-Ox_{(2)}$$
$$\overset{\bullet}{Red}_{(1)}-X-Ox_{(2)} \rightleftharpoons Ox_{(1)}-X-\overset{\bullet}{Red}_{(2)}$$
$$Ox_{(1)}-X-\overset{\bullet}{Red}_{(2)} \xrightarrow{k_2} Ox_{(1)}-X + \overset{\bullet}{Red}_{(2)}$$

and if the reverse redox path and the bridge breaking can occur at similar rates the actual bridging product will not build up any significant concentration but the kinetics will depart from the simple second-order rate law. Thus, in the reaction

$$Cr^{2+}aq + [Ru(NH_3)_5Cl]^{2+} \longrightarrow [Cr(H_2O)_5Cl]^{2+} + Ru(NH_3)_5OH_2]^{2+}$$

the rate becomes independent of the concentration of Cr^{2+} aq when this is present in sufficient excess. This limiting rate represents the aquation of the $[(NH_3)_5Ru^{II}ClCr^{III}(H_2O)_5]^{4+}$ complex.

9-6-4 Electron transfer to the bridge

When the bridging ligand is potentially reducible it is possible that the act of electron transfer is split into two steps and goes through an intermediate where the electron resides mainly on the bridge. For such a process to be observable it is necessary not only for the bridge to be able to accommodate the electron but also for the oxidant to be not over eager to take the electron. An example of this type of reduction is to be found in redox reactions with iso-nicotinamide as bridging ligand

$$(NH_3)_5CoN\!\!\!\bigcirc\!\!\!-C(=O)(NH_2) + Cr^{2+}aq \longrightarrow$$

$$Co^{2+}aq + 5NH_3 + N\!\!\!\bigcirc\!\!\!-C(=OCr(H_2O)_5^{3+})(NH_2)$$

$k = 17\cdot6\ M^{-1}s^{-1}\ (25°)$

or

$$(H_2O)_5CrN\!\!\!\bigcirc\!\!\!-C(=O)(NH_2)^{3+} + Cr^{2+}aq \longrightarrow$$

$$Cr^{2+}aq + N\!\!\!\bigcirc\!\!\!-C(=O\cdot Cr(H_2O)_5^{3+})(NH_2)$$

$k = 1\cdot8\ M^{-1}s^{-1}\ (25°)$

The reason for assigning the modified mechanism to this system is that, contrary to previous observations, the rate is not sensitive to the overall free energy change. [Usually the Co(III)+Cr(II) → Co(II)+Cr(III) reaction is some 10^5 to 10^7 times faster than the analogous Cr(III)+Cr(II) → Cr(II)+Cr(III) process.] With protonated fumarate as bridging ligand

$$R-O-C(=O)-C(H)=C(H)-C(OH)=O$$

the Cr^{2+}aq reduction of R = Co(NH$_3$)$_5$ is even slower (about half the rate) than the reduction of R = Cr(H$_2$O)$_5$. Whether or not the transfer of the electron to the bridge becomes a separate stage depends not only upon the nature of the reductant and the bridging ligand but also upon the oxidant. Thus

$$[(NH_3)_5RuN\text{—}C_6H_4\text{—}C(NH_2)=O]^{3+}$$

has similar oxidizing properties to the Co(III) analogue and yet its rate of reaction with Cr^{2+} aq is some 2×10^4 times faster and presumably utilizes a different mechanism. It has been suggested that the Co(III) and Cr(III) atoms allow the electron to reside on the bridging ligand because both accept the electron in an e_g orbital. This has σ-symmetry and therefore does not interact directly with the π-system of the ligand, which presumably accommodates the extra electron. The t_{2g}^5 Ru(III) accepts the electron into a t_{2g} orbital which has π-symmetry and can therefore interact directly with the π system of the bridging ligand.

9-7 Effect of the non-bridging ligands

The availability of substitutionally inert reagents has made it possible to study the effects of the nature and the position of the non-bridging ligands upon the rate of the redox reactions. The effect is quite marked and depends upon the nature of the species concerned. Most of the data available refers to the Co(III), Cr(III), and Ru(III) oxidants. For Co(III) complexes of the type $[Co\ en_2A\ Cl]^{n+}$, in their reaction with Fe^{2+}, an inner sphere mechanism can be demonstrated by the transient appearance of $FeCl^{2+}$. When A is *cis* to the chlorine the order of effectiveness is $NH_3 < -NCS < H_2O < Cl$, the rates covering the range from 1.8×10^{-5} to $1.6 \times 10^{-3} M^{-1} s^{-1}$ at 25°. When A is *trans* to the bridge the effect is much more marked, the sequence being $NH_3 < -NCS < Cl < N_3 < H_2O$ with the rates covering the range 6.6×10^{-5} to $2.4 \times 10^{-1} M^{-1} s^{-1}$ at 25°. In a less systematic way similar reactivity patterns have been observed for the reactions of Cr(III) oxidants. It has been suggested that the effectiveness of the non-bridging ligand is related to its ligand field strength—the greater the ligand field strength, the less reactive the complex—but it is more likely, in so far as the distinction can be made, that the σ-bonding ability of the ligand is more relevant. Both Co(III) and Cr(III) are reduced by taking an electron into an e_g orbital. Orgel suggested that this would correspond to the d_{z^2} orbital and the electron would therefore interact most with the bridging ligand and the one *trans* to it. In more general terms we can say that, as a consequence of the Franck-Condon restrictions, the oxidant must adjust its bond lengths before it can receive the electron. This will mainly cause a stretching of the M-bridge bond and its *trans* partner so that a strongly bonded ligand trans to the bridge would produce a less reactive oxidant than a weakly bound ligand. The cis-[Co en$_2$A Cl]$^{n+}$ oxidants have the same ligand trans to the chloride bridge and variation of A has a smaller, but still significant, effect on the reactivity. It is of interest to note that these reactivity patterns also apply when these complexes are reduced with an outer sphere mechanism. On the other hand, the reactions of Ru(III) complexes with V^{2+}, which have an outer sphere mechanism, do not conform to this pattern and ammonia ligands give rise to slightly better oxidants than the far less strongly bound water, c.f. Ru(H$_2$O)$_5$Cl^{2+} (1.9×10^3 M^{-1}s^{-1}) with Ru(NH$_3$)$_5$Cl^{2+} (3.0×10^3 M^{-1}s^{-1}) and Ru(NH$_3$)$_5$Cl^{2+} with trans-[Ru(NH$_3$)$_4$H$_2$O Cl]$^{2+}$ (1.4×10^3 M^{-1}s^{-1}). Here, it is suggested that, since the accepting t$_{2g}$ orbital has π-symmetry it is the π- and not the σ-bonding properties of the non-bridging ligands that are important.

9-8 Multiple bridging

So far we have only discussed reactions in which there is a single bridge, that is, the bridge occupies one coordination site on the oxidant and one on the reductant, and where bridge transfer takes place only one ligand is transferred. This is true even when multiple bridges might be expected from the known coordination behaviour. Thus in the reaction between Cr^{2+}aq and cis-[Co en$_2$(OH)$_2$]$^+$ only one oxygen is transferred from cobalt to chromium in spite of the great tendency of the latter to form cis diol bridges. In a similar way Cr^{2+}aq reacts with cis-[CrCl$_2$(H$_2$O)$_4$]$^+$ to give [CrCl(H$_2$O)$_5$]$^{2+}$, indeed the rate of the redox process is equal to the rate of the chloride release. One might ask why the system should wait for a second bridge to be formed before electron transfer takes place if one is enough. However, there is evidence that multiple bridges can form and even be transferred. The first example reported was deduced from the reaction between Cr^{2+}aq and cis-[Cr(N$_3$)$_2$(H$_2$O)$_4$]$^+$ where, in contrast to the reaction of the analogous dichloro complex, the rate of electron transfer as measured by ^{51}Cr exchange between the two species is considerably faster than the rate of azide release. This indicates that a double bridge must form and transfer

$$(H_2O)_6Cr^{2+} + \begin{array}{c} N=N=N \\ \diagdown \\ \overset{*}{C}r(H_2O)_4^+ \\ \diagup \\ N=N=N \end{array} \rightleftharpoons (H_2O)_4Cr \begin{array}{c} N=N=N \\ \diagdown \\ \diagup \\ N=N=N \end{array} \overset{*}{C}r(H_2O)_4^{3+}$$

$$+ 2H_2O \updownarrow$$

$$(H_2O)_4Cr \begin{array}{c} N=N=N \\ \diagup \\ \diagdown \\ N=N=N \end{array} + \overset{*}{C}r(H_2O)_6^{2+}$$

Later it was shown that cis-$[Cr(N_3)_2(H_2O)_4]^+$ is among the products of the reaction of Cr^{2+}aq with cis-$[Co\ en_2(N_3)_2]^+$. Certain dicarboxylato species, for example, cis-$[Co\ en_2(HCOO)_2]^+$ and cis-$[Co(NH_3)_4(CH_3COO)_2]^+$ also transfer both ligands and it was suggested at one time that when $CoEDTA^-$ was reduced by Cr^{2+}aq *three* of the carboxylates of the hexadentate were transferred.

9-9 Differentiation between outer sphere and inner sphere mechanisms

It is now reasonably clear that the outer sphere mechanism is open to all redox reactions and may even be the favoured path when a bridging system is available. Thus, while Cr^{2+}aq can rapidly accept a bridging ligand and oxidize by way of an inner sphere mechanism and seems to prefer this, it can also act as an outer sphere reductant not only when no bridge is presented, as it does reluctantly with $Co(NH_3)_6^{3+}$, but also when a perfectly good bridge is present as with $IrCl_6^{2-}$ (here parallel outer and inner sphere paths have been demonstrated). $Cr(bipy)_3^{2+}$, on the other hand, reacts as an outer sphere reductant. $Co(CN)_5^{3-}$ invariably appears to behave as an inner sphere reductant and outer sphere processes in this system have a kinetic rate law of the form Rate = k[oxidant][$Co(CN)_5^{3-}$][CN^-], and yield as product, $Co(CN)_6^{3-}$. Here it seems that the thermodynamically very unstable $Co(CN)_6^{4-}$ is a highly effective outer sphere reductant. $Co(NH_3)_5X^{n+}$, while providing a perfectly respectable bridge in X can react either way, depending upon the requirements of the reductant and the bridging effectiveness of X, and the mode of reaction can be deduced from the reactivity pattern. Thus, with Cr^{2+}aq the reactions are generally inner sphere and the rate for X = H_2O (5×10^{-1} M^{-1} s^{-1}, 25°) is much less than for X = OH($1 \cdot 5 \times 10^6$ M^{-1} s^{-1}, 25°). For the outer sphere reaction with $Cr(bipy)_3^{2+}$ X = H_2O and X = OH have similar rates (5×10^4, and 3×10^4 M^{-1} s^{-1}, 25°, respectively). The corresponding rate constants for the reaction with $Ru(NH_3)_6^{2+}$, another outer sphere reductant, are $3 \cdot 0$ and 4×10^{-2} M^{-1}, s^{-1}, 25°, respectively. With $V(H_2O)_6^{2+}$ as reductant, we can make use of the fact that the rate of the inner sphere mechanism is controlled by the rate of substitution in the coordination shell of $V(H_2O)_6^{2+}$ and any redox reaction whose rate is significantly in excess of this must occur by an outer sphere mechanism. Reductions by $Ru(NH_3)_6^{2+}$ which is substitutionally inert must, if fast enough, be outer sphere.

Therefore by making use of redox systems of suitably controlled substitution lability it is possible to document authentic outer sphere mechanisms where electron transfer is shown to take place without disturbance of the coordination shells of the reactants and list their reactivity behaviour patterns. In the same way authentic inner sphere mechanisms can be identified by the transfer of the bridging ligands (generally from oxidant to reductant) and the pattern of behaviour identified. Taking these behaviour patterns as mechanistic criteria it has then been possible to examine substitutionally labile systems in the hope of identifying the nature of the redox mechanism. In this way, reductions by Eu^{2+}aq and Fe^{2+}aq are thought to go by inner sphere mechanisms when bridges are available, and recently, transient $Fe^{III}X\,aq^{2+}$ intermediates have been identified.

9-10 Number of electrons transferred

Until now, in order to introduce as few complications as possible, discussion has been restricted to processes in which a single electron has been transferred from the reductant to the oxidant. Most of the examples considered involved complexes of transition metals whose stable oxidation states could be separated by one unit. However, the elements of the P-block and even the transition elements in covalent combination do not usually form stable compounds with an odd number of electrons and the stable oxidation states are separated by two units, as for example Sn(II) and Sn(IV); Tl(I) and Tl(III); P(III) and P(V). This then leads to the question: Is it possible to transfer more than one electron in a single act of oxidation–reduction? For an outer sphere reaction, the Franck–Condon restriction (that is, the matching of the energies of the oxidant and the reductant before electron transfer) will be far more serious if two electrons have to be transferred and the higher activation energies and the lower probability of a successful redox encounter will both serve to make such a process unlikely. In solution, the restrictions might be apparently mitigated by the fact that the reagents can stay together (in a solvation trap) long enough to participate in two separate acts of electron transfer. If the reactive intermediates do not have sufficient time to escape and be trapped or otherwise detected it is likely that this type of process cannot be genuinely distinguished from a synchronous two-electron transfer process. For an inner sphere mechanism the bridge can hold together long enough for more than one electron to transfer, one after the other. In principle, it is possible to distinguish two extreme situations, which may not necessarily be as different as they seem. The one extreme is completely identical to the inner sphere processes already mentioned—that is, the bridging ligand, with a spare pair of electrons, acts as a Lewis base to form the bridge with the other component acting as a Lewis acid. A good example of this is to be found in the Pt(II) catalysed chloride exchange of, say, *trans*-Pt en$_2$Cl$_2^{2+}$ which involves a two-electron transfer accompanied by chloride bridge transfer between a five-coordinate Pt(II) species and a six-coordinate Pt(IV) species:

$$Pt\ en_2^{2+} + \overset{*}{Cl}^- \rightleftharpoons Pt\ en_2\overset{*}{Cl}^+ \quad (fast)$$

$$\overset{*}{Cl}-\overset{en}{\underset{en}{Pt}}{}^{II+}:Cl-\overset{en}{\underset{en}{Pt}}{}^{IV}-Cl^{2+} \rightleftharpoons \overset{*}{Cl}-\overset{en}{\underset{en}{Pt}}{}^{II}-Cl-\overset{en}{\underset{en}{Pt}}{}^{IV}-Cl^{3+}$$

$$\updownarrow$$

$$\overset{*}{Cl}-\overset{en}{\underset{en}{Pt}}{}^{IV}-Cl^{2+} + \overset{en}{\underset{en}{Pt}}{}^{II}-Cl^+ \rightleftharpoons \overset{*}{Cl}-\overset{en}{\underset{en}{Pt}}{}^{IV}-Cl-\overset{en}{\underset{en}{Pt}}{}^{II}-Cl^{3+}$$

It is interesting to note that the four-coordinate planar Pt^{II} species is ineffective in this redox reaction. The Franck–Condon restriction requires that both metal atoms are equivalent for electron transfer and this means that either the reductant must increase its coordination number to five or the oxidant must decrease its coordination number to five before the bridge can be formed and, in this case, the former way out is adopted. The other extreme effectively requires the bridging atom to act as the Lewis acid and the reductant to act as the Lewis base. In this type of mechanism a pair of non-bonding electrons on the reductant become bonding. Our electron accounting system then tells us that a two-electron redox process has taken place.

Thus,

$$O_3\overset{IV}{S}{:}^{2-} + O\overset{V}{Cl}O_2^- \longrightarrow O_3S{:}OClO_2^{3-} \longrightarrow O_3\overset{VI}{S}O^{2-} + {:}\overset{III}{Cl}O_2^-$$

can be looked upon as a nucleophilic displacement of ClO_2^- from oxygen. This example may be somewhat equivocal but few people doubt that the reaction:

$$py{\cdots}\overset{dmg}{\underset{dmg}{Co}}{:}^{I-} + H_3C-\overset{dmg}{\underset{dmg}{Co}}{}^{III}-py \longrightarrow py-\overset{dmg}{\underset{dmg}{Co}}{}^{III}-CH_3 + {:}\overset{dmg}{\underset{dmg}{Co}}{\cdots}py^-$$

is nothing more than a nucleophilic attack on the carbon by the very strong Co(I) nucleophile. In this case, the methyl group has no great tendency to act as a bridging ligand (an electron deficient three-centre bond as in $Al_2(CH_3)_6$ seems rather far-fetched in this particular substitution). But if we are considering far-fetched explanations we ought not to dismiss the possibility that, in the $Pt^{II}en_2^{2+} + Cl^- + Pt\ en_2^{IV}Cl_2^{2+}$ reaction, the activation is an attack by the Pt(II) nucleophile (which is after all isoelectronic with the Co(I) species) upon the chloride of the Pt(IV) complex (presumably using an empty d orbital). This is, however, very unlikely since the nucleophilicity of such a transition metal ion is very sensitive to the ligand environment and, although the Pt(II) species can be a reasonable nucleophile when surrounded by 'softer' ligands, the above amine complex has little formal nucleophilic properties. Such reactions will be discussed in the next chapter.

9-11 Complementary and non-complementary reactions

As has just been pointed out, many redox couples differ by one unit of oxidation state, for example, Fe(II): Fe(III); Co(II): Co(III); Cr(II): Cr(III); Ru(II): Ru(III); while others differ by two units—Tl(I): Tl(III); Sn(II): Sn(IV); Rh(I): Rh(III); Pt(II): Pt(IV); Au(I): Au(III). If we consider a complete redox process which may or may not involve a single act of reaction, we can visualize two types of situation:

(i) The oxidant and the reductant change their oxidation states by an equal number of units. This is termed a **complementary** reaction.

(ii) The oxidant and the reductant change their oxidation states by a different number of units. Such reactions are termed, **non-complementary**.

Some examples are shown in Table 9-5.

Table 9-5

(a) *Complementary reactions*

$$Co(III) + Cr(II) \longrightarrow Co(II) + Cr(III)$$
$$Co(III) + Fe(II) \longrightarrow Co(II) + Fe(III)$$
$$V(II) + Tl(III) \longrightarrow V(IV) + Tl(I)$$
$$Sn(II) + Tl(III) \longrightarrow Sn(IV) + Tl(I)$$

(b) *Non-complementary reactions*

$$2Fe(III) + Sn(II) \longrightarrow 2Fe(II) + Sn(IV)$$
$$2Cr(II) + Tl(III) \longrightarrow 2Cr(III) + Tl(I)$$
$$2Cr(II) + Pt(IV) \longrightarrow 2Cr(III) + Pt(II)$$
$$Cr(VI) + 3Fe(II) \longrightarrow Cr(III) + 3Fe(III)$$
$$Mn(VII) + 5Fe(II) \longrightarrow Mn(II) + 5Fe(III)$$

It is clear that, in the case of the non-complementary reaction, it is going to be rather difficult to satisfy everybody in one go and one of the components at least will have to change by way of an unstable oxidation state. It is less obvious, but nevertheless possible, that in complementary reactions there may be reactive intermediates in unstable oxidation states.

Provided these intermediates can escape from their initial environment, they can be detected. The identification of the reactive intermediate generally provides the clue to the number of electrons transferred in the rate-determining step.

A number of techniques of general or specific application have been used.

9-11-1 Trapping

The reactive intermediate oxidation state may be especially partial to a particular reagent and be captured by it in preference to following the normal course of the original reaction. This will immediately show up as a change in the stoichiometry of the reaction as a result of adding the trapping reagent. Alternatively one can see whether or not this reagent is consumed. For example, it has been shown that Sn(III) reacts very rapidly with $Co(C_2O_4)_3^{3-}$ and causes its decomposition. Sn(II) interacts with the trap very much more slowly. Thus, if a reaction involving the Sn(II)—Sn(IV) couple is carried out in the presence of $Co(C_2O_4)_3^{3-}$ it is possible to state, by looking at what happens to the Co(III) complex, whether or not Sn(III) is formed as an intermediate and is left alone long enough for it to reduce the Co(III). Thus, if the reactions between Sn(II) and Cr(VI) or Mn(VII) are carried out in the presence of $Co(C_2O_4)_3^{3-}$ there is reduction of the cobalt(III) and Sn(III) must be postulated as an intermediate. On the other hand, $Co(C_2O_4)_3^{3-}$ is unaffected when it is present at the reaction between Sn(II) with Tl(III) or Hg(II) and one can assume that Sn(III) is either not formed (two-electron transfer) or else it never has a chance to reduce the Co(III).

9-11-2 Chemical behaviour

This heading includes a very large number of reactions that are completely specific to a particular reagent. For example, the reaction of $Co(NH_3)_5C_2O_4^+$ with a variety of oxidizing agents (oxidation of coordinated oxalate) can follow one of two paths depending upon the nature of the oxidant. In the first, the products are $Co(NH_3)_5H_2O^{3+} + 2CO_2$ and two equivalents of oxidant are consumed for each mole of cobalt complex destroyed. The second group of oxidants form $Co(H_2O)_6^{2+} + 5NH_3 + 2CO_2$ and only one equivalent of oxidant is consumed. It has been suggested that the first group of oxidants, for example, Cl_2, Br_2, Tl(III), Sn(IV), remove two electrons in one step.

$$[Co^{III}(NH_3)_5(C_2O_4^=)]^+ - 2e^- \longrightarrow [Co^{III}(NH_3)_5(C_2O_4)]^{3+}$$
$$\downarrow$$
$$[Co^{III}(NH_3)_5H_2O]^{3+} \xleftarrow{H_2O} [Co^{III}(NH_3)_5]^{3+} + 2CO_2$$

and the neutral C_2O_4 ligand splits into two carbon dioxide molecules and falls off the cobalt, its place being taken by water. The second group of oxidants, Fe(III), Ce(IV), remove just one electron and the $C_2O_4^-$ radical ion produced is unwilling to wait for a second visit by the oxidant and reduces the cobalt(III) to which it is coordinated.

$$[Co^{III}(NH_3)_5(C_2O_4^=)]^+ - e^- \longrightarrow [Co^{III}(NH_3)_5(C_2O_4^-)]^{2+}$$
$$\downarrow$$
$$Co^{2+}aq + 5NH_3 + 2CO_2 \longleftarrow [Co^{II}(NH_3)_5(C_2O_4)]^{2+}$$

The alternative paths for the oxidation of hydrazine:

$$N_2H_4 \longrightarrow N_2 + 4H^+ + 4e^-$$

and $N_2H_4 \longrightarrow NH_3 + \tfrac{1}{2}N_2 + H^+ + e^-$

relate to the oxidant used and whether or not it removes two electrons in the first encounter.

The substitutionally inert character of chromium(III) can be used to throw light upon the nature of the oxidant of chromium(II). A one-electron inner sphere, or even outer sphere, reaction will produce mononuclear species such as $Cr(H_2O)_5X^{2+}$ [or even $Cr(H_2O)_4X_2^+$ in the rare cases where a double bridge is transferred] or $Cr(H_2O)_6^{3+}$ and these are all readily identifiable. However, oxidation of Cr^{2+}aq by oxygen or Tl(III) gives the green diol bridged binuclear species

$$(H_2O)_4Cr\begin{matrix}\diagup OH \diagdown \\ \diagdown OH \diagup\end{matrix}Cr(H_2O)_4^{4+}$$

which is formed by the reaction

$$Cr(IV) + Cr(II) \longrightarrow [Cr(III)]_2$$

The formation of this product is a clear indication that somewhere in the reaction sequence there has been a two-electron oxidation of Cr(II) to Cr(IV). The reaction between Cr^{2+}aq and $Pt(NH_3)_5Cl^{3+}$ is even more complicated because, in addition to being a two-electron process, the chlorine is transferred and the $Cr^{IV}Cl^{3+}$ intermediate reacts with Cr^{II} to give a 50:50 mixture of $CrCl(H_2O)_5^{2+}$ and $Cr(H_2O)_6^{3+}$. Here the chloride bridged Cr_2Cl^{5+} dimer is substitutionally labile. The green diol dimer can be observed in the product when the reaction is carried out at higher pH.

9-11-3 Kinetic behaviour

If a highly reactive species is formed as an intermediate it is likely that it will not be greatly discriminating in its subsequent reaction and so the

148 Oxidation and reduction Ch. 9

process of its formation will be reversible. An alternative way of putting it would be to say that the slow step, leading to an unstable oxidation state, would often involve an overall gain in free energy and therefore be strongly reversible. Thus by adding the stable product of the first stage to the reaction mixture the formation of the reactive intermediates will be reversed and the rate of reaction decreased. An example could be the non-complementary reaction

$$2Fe(II) + Tl(III) \longrightarrow 2Fe(III) + Tl(I)$$

which could either go by a succession of one-electron transfers:

$$Fe(II) + Tl(III) \rightleftharpoons Fe(III) + Tl(II) \quad \text{slow} \tag{9-1}$$
$$Fe(II) + Tl(II) \longrightarrow Fe(III) + Tl(I) \quad \text{fast} \tag{9-2}$$

in which Tl(II) is the unstable intermediate; or else by a slow two-electron transfer reaction:

$$Fe(II) + Tl(III) \rightleftharpoons Fe(IV) + Tl(I) \quad \text{slow} \tag{9-3}$$
$$Fe(II) + Fe(IV) \longrightarrow 2Fe(III) \quad \text{fast} \tag{9-4}$$

in which Fe(IV) is the reactive intermediate. In the one-electron mechanism, the addition of excess Fe(III) ought to slow the reaction by returning some of the Tl(II) to the original Tl(III) while in the two-electron mechanism addition of excess Tl(I) should fulfil the same role by returning the Fe(IV) to Fe(III). Experiment shows that Fe(III) retards the reaction whereas Tl(I) does not, adequate evidence to demonstrate a one-electron transfer process with Tl(II) as intermediate. This is a reasonably common mode of reaction for Tl(III) with oxidizable transition metal ions.

9-12 Catalysis of non-complementary reactions

We have already seen how, in a simple non-complementary process, one of the components will be forced to adopt an unusual oxidation state in the first stage of reaction and that the energetics of the process may very well cause it to be extremely reversible and lead to a slow overall reaction. An opportunity is thus provided for catalysis in which the catalyst acts as mediator and avoids the need for the unstable oxidation state. Thus non-complementary reactions are generally very sensitive to the nature of the ligands present and indeed other metal ion impurities. Two major types of catalysis can be encountered.

9-12-1 Catalysis by potential bridging ligands
In this type of assistance, the one-electron component forms a bridged dimer in a pre-equilibrium and it is this that reacts with the two-electron component. This mechanism has been invoked for the reaction between

Co(II) and Pb(IV) in glacial acetic acid

$$2Co(II) \rightleftharpoons Co(II)_2$$
$$Co(II)_2 + Pb(IV) \longrightarrow 2Co(III) + Pb(II)$$

and provides a means whereby the termolecular process, as required by stoichiometry, is achieved without the handicap of a very low probability of a spontaneous triple collision.

9-12-2 Catalysis by another redox couple

If the redox couple of the catalyst can cope with both one- and two-electron transfers it can mediate effectively between the reagents of a non-complementary reaction.

An example of this can be found in our analytical experience. The peroxydisulphate ion is a very powerful oxidizing agent,

$$S_2O_8^{2-} + 2e^- \longrightarrow 2SO_4^{2-}$$

but if odd-electron species are not to be involved it is a *two*-electron oxidant. Oxidation of species that prefer *one*-electron changes can therefore be slow. In the oxidation of Cr(III) to Cr(VI) by $S_2O_8^{2-}$ (a well-known stage in the quantitative estimation of Cr(III)) $AgNO_3$ is added as a catalyst. It is likely that its role as catalyst comes from its ability to be oxidized to Ag(III) which then reacts with more Ag(I) to give Ag(II):

$$Ag(I) + S_2O_8^{2-} \longrightarrow Ag(III) + 2SO_4^{2-}$$
$$Ag(III) + Ag(I) \longrightarrow 2Ag(II) \qquad \text{fast}$$

This then acts as a one-electron oxidant:

$$Cr(III) + Ag(II) \longrightarrow Cr(IV) + Ag(I)$$
$$Cr(IV) + Ag(II) \longrightarrow Cr(V) + Ag(I)$$
$$Cr(V) + Ag(II) \longrightarrow Cr(VI) + Ag(I)$$

Catalysis by a second redox couple is not confined to non-complementary reactions and can occur simply because the rate of reaction of oxidant and reductant with the reduced and oxidized form of the catalyst is so much faster than their direct reaction.

Thus, in the reaction

$$Fe(III) + V(III) \longrightarrow V(IV) + Fe(II)$$

(which is not as simple as it looks) addition of Cu(II) speeds up the reaction and the catalysed path is independent of the concentration of Fe(III). It is found that the rate constant for the reaction

$$V(III) + Cu(II) \longrightarrow V(IV) + Cu(I)$$

is 12 times greater than that for the direct reaction and that the subsequent reaction

$$Cu(I) + Fe(III) \longrightarrow Cu(II) + Fe(II)$$

is very fast.

Problems

9-1 Under what conditions is $Co(CN)_6^{3-}$ the product of oxidation of solutions of Co(II) containing a sufficient excess of cyanide? What special feature enters the kinetic rate law?

9-2 Suggest a quantitative method of determining the relative abilities of a series of ligands X in $Co(NH_3)_5X^{n+}$ to act as bridging groups in the reduction by $Cr\ aq^{2+}$ that does not require the measurement of the rates of fast reactions?

9-3 Cobalt(II), in the presence of cyanide, reduces $[Co(NH_3)_5F]^{2+}$ and $[Co(NH_3)_5CH_3COO]^{2+}$ by an outer sphere mechanism and $[Co(NH_3)_5Cl]^{2+}$ and $[Co(NH_3)_5Br]^{2+}$ by an inner sphere mechanism. On what evidence is this statement based?

9-4 The reaction $Hg_2^{2+} + Tl^{3+} \longrightarrow 2Hg^{2+} + Tl^+$ obeys the rate law $-d[Tl^{3+}]/dt = k[Tl^{3+}][Hg_2^{2+}]/[Hg^{2+}]$. Write down a suitable mechanism and show how the rate law can be derived. Is this a one- or a two-electron process? [See A. M. Armstrong, J. Halpern, and W. C. E. Higginson, *J. Phys. Chem.*, 1956, **60**, 1661.

9-5 How would you show by kinetics that the reaction between Cr_{aq}^{2+} and Co_{aq}^{3+} went by direct electron transfer and did not involve a two-stage process of the type $Cr^{2+}aq \longrightarrow Cr^{3+}aq + e^-aq$, $e^-aq + Co^{3+}aq \longrightarrow Co^{2+}aq$?

Bibliography

Anbar, M., The reactions of hydrated electrons with inorganic compounds. *Quart. Revs.*, 1968, **22**, 578.

Earley, J. E., Non-bridging ligands in electron transfer reactions. *Prog. Inorg. Chem.* (Ed. J. O. Edwards), 1970, **13**, 243.

Halpern, J., Mechanisms of electron transfer and related processes in solution. *Quart. Revs.*, 1961, **15**, 207.

Hart, E. J. and M. Anbar. *The Hydrated Electron*. John Wiley (Interscience), New York, 1970.

Reynolds, W. L. and R. W. Lumry. *Mechanisms of Electron Transfer*. The Ronald Press Company, New York, 1966.

Ruff, I., The theory of thermal electron-transfer reactions in solution. *Quart. Revs.*, 1968, **22**, 199.

Sutin, N., Free energies, barriers, and reactivity patterns in oxidation-reduction reactions. *Acc. Chem. Res.*, 1968, **1**, 225.

Sykes, A. G., Further advances in the study of mechanisms of redox reactions. *Adv. Inorg. Chem. Radiochem.* (Ed. H. J. Emeléus and A. G. Sharpe), 1967, **10**, 153.

Taube, H. and E. S. Gould. Organic molecules as bridging groups in electron transfer reactions. *Acc. Chem. Res.*, 1969, **2**, 321.

Taube, H., *Electron Transfer Reactions of Complex Ions in Solution*. Academic Press, New York, 1970.

Redox addition, elimination, and substitution

10-1 Introduction

In previous chapters it was convenient to concentrate upon those reactions in which changes in the coordination shell could be considered quite independently from changes in the oxidation state, or vice versa. Sometimes the distinction was not clear-cut as in the case of the inner sphere redox process in which the formation and breaking of the necessary bridge were acts of ligand substitution, but even there it was usually possible to consider the substitution and redox parts as separate aspects of a multi-stage process. In this chapter we will see how the principles that have been established in the 'simple' reactions can be applied to systems where the two aspects are completely interlinked. A major factor that underlies these redox-controlled changes in the coordination shell is the relationship between electron configuration and coordination number. In any general discussion of the problems it could be stated that at one extreme, in the more electrovalent chemistry of the D-block elements, coordination numbers are determined by the interplay of factors such as charge, ligand electronegativity, size, packing, and other steric effects. At the other extreme, in the more covalent chemistry, the coordination number is governed to a great extent by the orbitals that are available and the '18-electron' rule functions satisfactorily (assuming that non-bonding d electrons pair off wherever possible). Steric effects have to be fairly drastic to modify these rules. In the middle region between these extremes the symmetry relationships between the orbitals used and the coordination geometry become important, and the effect of the ligands upon the relative energies of the electron in these orbitals will control the coordination geometry and the spin multiplicity.

Any discussion of reactions in which coordination shell changes and oxidation state changes are completely intertwined will focus mainly upon those areas where the '18-electron' rule tends to operate and at this stage of development virtually all of the information comes from the d^6, d^8, and d^{10} configurations of the elements in the second and third row of the D-block. The relationship between electron configuration, oxidation state, and coordination number is set out in Table 10-1 where it will be seen that on going from d^6 to d^8 to d^{10} configurations, the predicted

Table 10-1 Electron configuration, oxidation state, and coordination number for second and third row D-Block elements[a] [b]

Configuration	Examples			Coordination numbers
d^{10}	Rh(-I) Ir(-I)	Pd(0) Pt(0)	Au(I)	2, 3, $\boxed{4}$[c]
d^8	Ru(0): Os(0)	Rh(I) Ir(I)	Pd(II) Pt(II)	4, $\boxed{5}$
d^6	Ru(II) Os(II)	Rh(III) Ir(III)	Pd(IV) Pt(IV)	$\boxed{6}$

[a] Mononuclear complexes with unequivocal odd electron configurations are rare.
[b] Cobalt and nickel will copy their congeners in certain covalent complexes.
[c] The coordination number predicted by '18-electron' rule is boxed.

coordination number decreases from six to five to four, but coordination unsaturation becomes more and more common. This is, in fact, a key to the types of reactions that complexes of atoms with such electron configurations undergo. The d^4, d^2, and d^0 configurations have not been included in this discussion and coordination unsaturation (with respect to the 18 electron rule) becomes the norm. A few examples involving d^4 systems [Mo(II), W(II); Tc(III), Re(III); Ru(IV), Os(IV)] are now available and the numbers will certainly increase as more of the right sort of reactions are investigated.

10-2 Oxidative addition

10-2-1 Nature of the reaction

If we ignore the rather obvious halogen oxidation reactions of the type

$$PtCl_4^{2-} + Cl_2 \longrightarrow PtCl_6^{2-}$$

or $Au(CN)_2^- + Cl_2 \longrightarrow AuCl_2(CN)_2^-$

which are perfectly acceptable examples of $d^8 \to d^6$ and $d^{10} \to d^8$ oxidative additions, the first compound in which the extent of the applicability of this type of reaction was realized was trans-[Ir(CO)Cl(PPh$_3$)$_2$]. Such was the interest in the behaviour of this compound that it was, and still is, referred to as Vaska's compound after the name of its developer, a singular compliment in inorganic chemistry where the cult of the personality is not encouraged.

The characteristic reaction of Vaska's compound is the following:

$$[Ir^I(CO)Cl(PPh_3)_2] + XY \longrightarrow [Ir^{III}(CO)Cl(PPh_3)(X)(Y)]$$

where the X—Y molecule has split to form two ligands and the four-coordinate planar Ir^I has been oxidized to six-coordinate octahedral

Ir^{III}. Examples are to be found for XY = H_2, CH_3—I, HCl, Br_2, I_2 Cl—HgCl, CH_3CO—Cl, R_3Si—H, and so on. In many cases the Ir^{III} product is unstable and undergoes further reaction so that the general complexity of the overall process can be very great and considerable unravelling is required. Analogues of Vaska's compound have been developed but as a rule the behaviour is extremely sensitive to the choice of the ligand, thus trans-[Rh(CO)Cl(PPh$_3$)$_2$] is far less effective than the Ir analogue, but RhCl(PPh$_3$)$_3$ makes up for it.

The $d^{10} \rightarrow d^8$ change is best characterized by Pt(PPh$_3$)$_4$ which will react with a number of molecules of the X—Y type to form either d^{10} adducts such as Pt(PPh$_3$)$_2$(XY) or d^8 complexes of the form trans-[Pt(PPh$_3$)$_2$(X)(Y)]. At first sight, it might seem that there should be no difficulty in distinguishing between adduct formation and oxidation but, when X and Y are multiply bonded, as in C_2H_4, $C_2(CF_3)_4$, O_2, and so on, it is necessary to look at the structure and dimensions of the product before reaching any conclusion. Pd(PPh$_3$)$_4$, Ni[P(OEt)$_3$]$_4$ are other useful reagents for $d^{10} \rightarrow d^8$ oxidative additions.

Although oxidative addition is usually associated with two- or sometimes one-electron changes, the apparent complications may arise when 'non-innocent' ligands are present in the coordination sphere. Thus, Ir(NO)(PPh$_3$)$_3$ is generally represented as a complex of d^{10} Ir(—I) in which the linear Ir—N≡O bond suggests NO^+ as the structure of the ligand (isoelectronic with CO). This species readily undergoes oxidative addition with XY = CH_3I, Cl_2 and so on, to give five-coordinate IrNO(PPh$_3$)$_2$(X)(Y) where the bent (123°)

$$Ir-\overset{..}{N}\overset{\displaystyle O}{\diagup\!\!\!\!\diagup}$$

bond angle suggests that the ligand is NO^- and that, in spite of the five-coordination, the iridium is in +III oxidation state with a d^6 configuration. A four-electron oxidation is therefore formally proposed.

10-2-2 Mechanism of oxidative addition

Most of the work published in the field of oxidative addition has been preparative in the sense that the emphasis has always been upon the relationships between reagents and products and the way in which the nature of the ligands and the reaction centre affects this. As we have seen in earlier chapters, this type of approach while being extremely valuable is also extremely dangerous because it is not easy to say whether or not the product is the result of a single act of reaction or the consequence of a large number of reaction steps. It might even be a 'thermodynamically controlled' product, in which case it will provide misleading evidence of mechanism. This is especially true in reversible oxidative addition processes that lead effectively to ligand exchange or substitution.

In recent years, kinetic measurements as a means of investigating mechanism have been applied to a wide range of oxidative additions. In general, such measurements are by no means easy because the reactions are often fast and, by their very nature, the reagents are sensitive to the 'small molecules of the environment', O_2, H_2O, or even N_2. The pattern that emerges is relatively simple. Oxidative additions to compounds such as trans-$[IrCl(CO)(PPh_3)_2]$ follow simple second-order rate laws of the type Rate = $k_2[Ir(I)][XY]$. This behaviour is typical of most oxidative additions where the compound that is being oxidized is coordinatively unsaturated. The rate is very sensitive to the nature of XY and to the nature of the complex but, if we restrict consideration to a series of complexes of the type trans-$[IrX(CO)L_2]$ where X is a halogen and L is a phosphorus donor ligand, then the reactivity, while decreasing along the sequence X = Cl > Br > I, is not particularly sensitive to changes in the nature of X or L. Under certain circumstances, coordinately saturated species such as the five-coordinate cobalt(I) species, $[Co(dmg)_2py]^-$, react with simple second-order kinetics, but in this case only one ligand can be attached in order to achieve coordination saturation in the d^6 product and the reaction takes the form

$$[Co(dmg)_2py]^- + XY \longrightarrow [Co(dmg)_2py.X] + Y^-$$

A more typical behaviour, however, is that of $[IrH(CO)(PPh_3)_3]$ whose kinetics of oxidative addition take the form

$$\text{Rate} = \frac{k_1 k_2 [Ir(I)][XY]}{k_{-1}[PPh_3] + k_2[XY]}$$

and indicate a mechanism in which the coordinatively saturated substrate must first lose a ligand and become unsaturated before it can undergo oxidative addition:

$$[Ir(H)(CO)(PPh_3)_3] \underset{k_{-1}}{\overset{k_1}{\rightleftharpoons}} [Ir(H)(CO)(PPh_3)_2] + PPh_3$$

$$[Ir(H)(CO)(PPh_3)_2] + XY \overset{k_2}{\longrightarrow} [Ir(H)(X)(Y)(CO)(PPh_3)_2]$$

Similar pre-dissociation is required for the reactions of $Pt(PPh_3)_4$, $Pd(PPh_3)_4$, and $NiH[P(OMe)_3]_4^+$, where oxidative addition is retarded by addition of extra phosphine or phosphite ligand. The point to bear in mind when discussing oxidative addition with coordinatively saturated substrates is that, for a two-electron oxidation, the 'ideal' coordination number increases by one. The oxidant X—Y ideally provides two ligands so that, if the substrate is coordinatively saturated, some ligands (that is, from the substrate) or potential ligand (from X—Y) must be released either before or during (possibly even after) the act of oxidative addition.

There are two clear-cut extreme mechanisms. In the first, which might be termed 'addition-rearrangement', the molecule X—Y attaches itself to the reaction centre and then simultaneously or subsequently undergoes a redistribution of electrons whereby the X—Y bond is broken and M—X and M—Y bonds are formed. A three-centre transition state is likely for this process and can be formulated as

$$L_nM + XY \longrightarrow L_nM(XY) \longrightarrow L_n M \begin{matrix} X \\ \vdots \\ Y \end{matrix} \longrightarrow L_nM \begin{matrix} X \\ \diagdown \\ Y \end{matrix}$$

$$d^n \qquad\qquad d^n \qquad\qquad\qquad\qquad d^{(n-2)}$$

This type of reaction has a number of characteristic features:

(i) It requires coordination unsaturation in the reduced form, otherwise the oxidized product will have a coordination number that is greater than that predicted by the 18-electron rule.

(ii) It is most common when neither X nor Y is strongly electronegative.

(iii) The addition is stereospecific and *cis*, that is X and Y are adjacent in the oxidized form.

(iv) If X or Y are asymmetric the reaction will proceed with retention of configuration about these centres.

Typical of this type of oxidative addition are the reactions where X—Y = H_2, H—CRR'R", H—SiRR'R". A major piece of indirect evidence in favour of this type of mechanism comes from the study of adducts which are unable, for one reason or another, to surmount the barrier of the subsequent rearrangement. Thus, $Pt(PPh_3)_4$ reacts readily and reversibly with ethylene to form $Pt(PPh_3)_2C_2H_4$ which is usually formulated as a π-olefin complex of the d^{10} Pt(0), whereas the reaction with $C_2(CF_3)_4$ leads to an apparently analogous species where the strongly electronegative substituents cause the back donation from the metal to the ligand to be sufficient to warrant the description in terms of oxidation to Pt(II) and the formation of a 'metallo-cyclopropane'

ring. By varying the substituents the whole range of behaviour can be covered indicating that there is no clear-cut division between addition and oxidative addition. A similar case is to be found in the addition of oxygen to these species. $Pt(PPh_3)_4$ reacts with oxygen to give $Pt(PPh_3)_2O_2$ irreversibly and it is interesting to note that the most effective reacting species in solution is $Pt(PPh_3)_3$ and not $Pt(PPh_3)_2$, as might have been expected from the nature of the product. Vaska's complex adds oxygen reversibly to form $IrClCO(PPh_3)_2(O_2)$ where an O—O bond is retained (= 1·30 Å) but it is of considerable interest to note that changing Cl to I removes the reversibility of the process and in $IrICO(PPh_3)_2O_2$, O—O = 1·51 Å. The dimensions of the coordinated dioxygen are thus extremely sensitive to the nature of the centre to which it is coordinated and one can conclude again that the borderline between addition and oxidative addition is very finely balanced. The change-over to a clear-cut oxidative addition with these multiply bonded X—Y species requires some further reaction. Thus, $Pt(PPh_3)O_2$ will react with SO_2 to form

$$(PPh_3)_2Pt \underset{O}{\overset{O}{\diagdown\!\!\!\diagup}} SO_2$$

which is an undisputable chelated sulphato complex of Pt(II). CO will form

$$(PPh_3)_2Pt \underset{O}{\overset{O}{\diagdown\!\!\!\diagup}} C{=}O$$

and N_2O_4 forms $Pt(PPh_3)_2(ONO_2)_2$. A whole range of reactions can cause the π-bonded olefins to form compounds in which there is a metal-carbon σ-bond and a net oxidation of the metal. These reactions form an important part of many organo-metallic catalysis sequences and will be considered more fully in the next chapter.

The other extreme mechanism for oxidative addition has already been mentioned briefly in a previous chapter and might be termed 'nucleophilic substitution'. In this case, an electron pair, previously non-bonding, is brought into the bonding scheme with the result that the metal effectively undergoes a two-electron oxidation. The metal ion therefore behaves as an electron pair donor and functions as a nucleophile in attacking a suitable reaction centre. A typical reaction in this category would be between the five-coordinate cobalt(I) species CoL_4A^-, where L_4 can be two dimethylglyoximes or a range of quadridentate garland or macrocyclic ligands and A is a harmless neutral ligand (for example pyridine) occupying the fifth position. All of these complexes have three features in common:

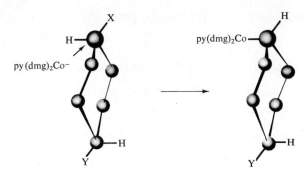

Fig. 10 – 1 Alkylation of Co^I (dmg)$_2$py$^-$ by 1,4-disubstituted cyclohexanes (X = Y = Br). Note the inversion at carbon

(i) The quadridentate or bis bidentate ligand system is conjugated.

(ii) The four donors remain in a planar arrangement.

(iii) They are some of the most powerful nucleophiles known with respect to their reactions with aliphatic carbon.

The interest in this type of reaction is considerable because Vitamin B_{12s} falls into this category and many of the complexes are looked upon as models for this particular vitamin system. These species are all co-ordinately saturated and the oxidative addition can be looked upon as

$$A-Co^I(L_4): + CH_3X \longrightarrow A-Co^{III}(L_4)-CH_3 + X^-$$
$$d^8 \qquad\qquad\qquad d^6$$

These reactions are typical S_N2 processes and have been shown to take place with inversion at carbon, most conveniently by using 1,2- or 1,4-disubstituted cyclohexanes and identifying the products by their ^{1}H.n.m.r. spectra (Fig. 10-1). The Rh(I) analogues behave in a similar way. Other coordinatively saturated d^8 complexes undergo similar reactions, for example, $Fe(\pi-C_5H_5)(CO)_2^-$, $Ru(\pi-C_5H_5)(CO)_2^-$, $Mn(CO)_5^-$, $Re(CO)_5^-$, and the reactivity is very sensitive to the nature of the reaction centre. The d^{10} $Co(CO)_4^-$ ion can act in a similar way, but less effectively. The two-electron oxidations of these coordinately saturated species lead to an increase of only one in the maximum coordination number and this is satisfied by the fragment 'X'. The rest of the oxidant is expelled as 'Y$^-$'. These reactions are therefore polar in character and the rates are very sensitive to the solvating properties of the solvent. Quite often the product Y$^-$ is an effective nucleophile and can displace a ligand from the initial oxidized product. This is commonly found in halogen oxidative addition of the type

$$\text{Os}^0(CO)_3(PPh_3)_2 + X_2 \longrightarrow [\text{Os}^{II}X(CO)_3(PPh_3)_2]^+ + X^-$$
$$d^8 \qquad\qquad\qquad\qquad\qquad d^6$$

$$[\text{OsX}(CO)_3(PPh_3)_2]^+ + X^- \longrightarrow \text{OsX}_2(CO)_2(PPh_3)_2 + CO$$

This can also be observed in a $d^6 \to d^4$ oxidative addition, where the ordination number increases from six to seven.

$$\text{Mo}^0(CO)_3 \text{ triars} + Br_2 \longrightarrow \text{Mo}^{II}Br(CO)_3 \text{ triars}^+ + Br^-$$
$$d^6 \qquad\qquad\qquad\qquad\qquad d^4$$

$$\text{MoBr}(CO)_3 \text{ triars}^+ + Br^- \longrightarrow \text{MoBr}_2(CO)_2 \text{ triars} + CO$$

(*triars* = bis(*o*-dimethylarsinophenyl)methylarsine, a terdentate ligand).

A considerable amount of work has been carried out on the oxidative addition reactions of coordinatively unsaturated d^8 species by XY reagents in which at least one component is sufficiently electronegative to favour extensive charge separation in the transition state and the mechanism approaches, at least, the limiting nucleophilic mechanism. What would be the possible consequences of such a mechanism when the reagent is coordinately unsaturated? Obviously the product too will be coordinately unsaturated and in the case of $d^8 \to d^6$ oxidative additions, it will be most unhappy in this condition and will take a ligand from wherever it can get it. A typical reaction of this type is

$$\text{Pt}^{II}(CN)_4^{2-} + Br_2 \longrightarrow \text{Pt}^{IV}(CN)_4BrH_2O^- + Br^-$$
$$d^8 \qquad\qquad\qquad\qquad d^6$$

where the coordinatively unsaturated $Pt(CN)_4Br^-$ produced initially takes up an adjacent water molecule either during or immediately after its formation. It would be tempting to consider the reactions of the analogues of Vaska's compound, *trans*-IrY(CO)(PR$_3$)$_2$ with alkyl halides as being nucleophilic substitutions with an unstable, coordinatively unsaturated, initial product, that is,

$$\textit{trans-}IrY(CO)(PR_3)_2 + CH_3X \longrightarrow [IrY(CH_3)(CO)(PR_3)_2]^+ + X^-$$
$$[IrY(CH_3)(CO)(PR_3)_2]^+ + X^- \longrightarrow IrXY(CH_3)(CO)(PR_3)_2$$

There is no question that in the sensitivity of the rates to the polarity of the solvent and the large negative entropies of activation these reactions resemble those of the coordinately saturated species and a considerable amount of charge separation in the transition state is indicated. However, the steric course of the addition indicates that the reaction might be more complicated than this and a considerable body of opinion exists in favour of a synchronous addition-rearrangement for this process.

The reaction of CH_3X with $[IrY(CO)(PR_3)_2]$, where X and Y are halogens, gives specifically *trans* addition when carried out in a poor ionizing solvent, such as benzene, or even under heterogeneous conditions as in the gas-solid reaction:

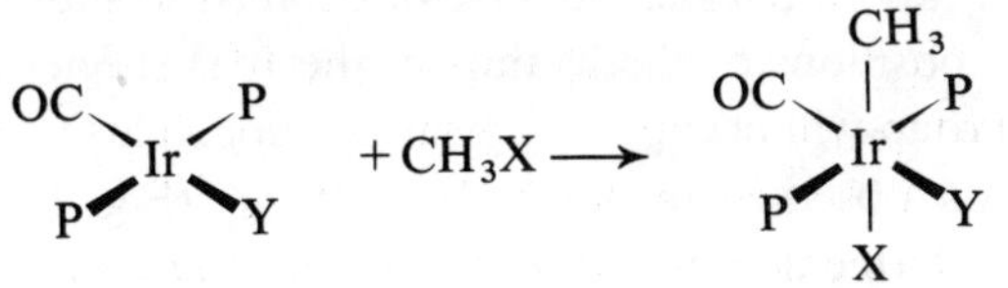

$$\text{OC}\diagdown\text{Ir}\diagup\text{P} + \text{CH}_3\text{X} \longrightarrow \text{OC}\diagdown\text{Ir}\diagup\text{P}$$
$$\text{P}\diagup\quad\diagdown\text{Y} \qquad\qquad \text{P}\diagup |\diagdown\text{Y}$$
$$\qquad\qquad\qquad\qquad\qquad\qquad\quad\text{X}$$

(with CH$_3$ above Ir in product)

but in more ionizing solvents, such as methanol, the process is far less stereospecific and a range of products are obtained. Some of these can only be accounted for in terms of prior or subsequent side reactions. If the mechanism is essentially nucleophilic with formation of the five-coordinate d^6 IrY(CH$_3$)(CO)(PPh$_3$)$_2^+$ and X$^-$ then we are faced, stereochemically, with a situation identical with that encountered in substitution reactions with an I_d or a D mechanism. The intermediate may well

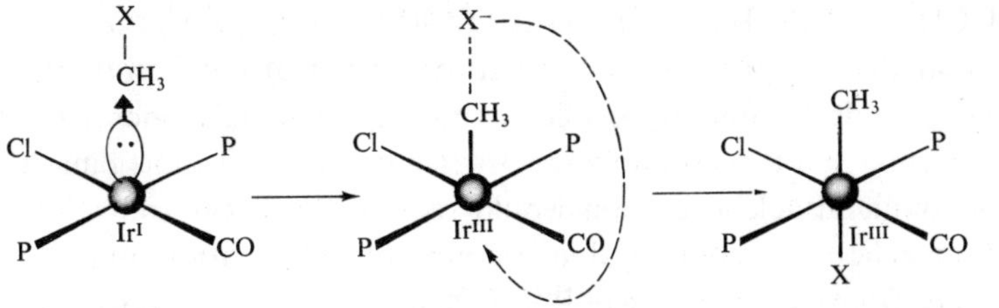

(a) *Trans* addition retaining original disposition of ligands attached to Ir$^\text{I}$—square pyramidal intermediate

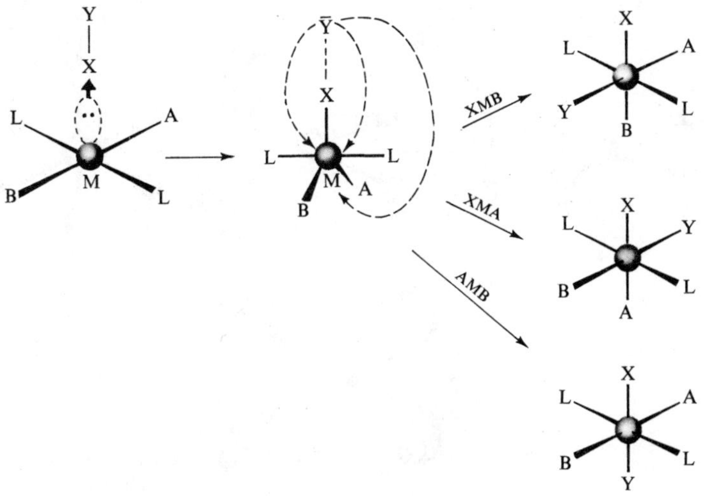

(b) Addition by way of a trigonal bipyramidal form giving *cis* as well as *trans* addition

Fig. 10 – 2

have been generated in a new way but its consumption will be governed by the same rules. A square pyramidal form might have been expected (in view of what we know about the steric course of octahedral substitution) and, if the relative positions of the ligands in the Ir(I) species are unchanged, the methyl group will occupy an apical position. This would lead to the *trans* addition observed in non-polar solvents (Fig. 10-2). A trigonal bipyramidal intermediate might allow *cis* and *trans* addition (Fig. 10-2) but it is not clear why this should be favoured in a polar solvent. An alternative explanation is that the lifetime of the square pyramidal intermediate is enhanced by solvation in polar solvents and it has enough time to undergo pseudorotation. It might seem indisputable that the observation of *trans*-addition is incompatible with the addition-rearrangement mechanism and its three-centre transition state, but it has been suggested that the orbital symmetry properties would not preclude *trans* addition from a three-centre transition state (Fig. 10-3). The main reason for this suggestion was the observation that oxidative addition of $CH_3CHBrCOOC_2H_5$ to *trans*-$[Ir(CO)Cl(PPh_2Me)_2]$ gave *trans* addition but that the reaction took place with retention of configuration at the carbon. While it is not impossible that the experimental evidence has been misinterpreted it is reasonably safe to assume that the simple bimolecular nucleophilic attack at carbon would require the linear transition state and inversion. A considerable amount of information ought to be obtained from a comparison of R_3C-X and R_3Si-X oxidative additions because, as we saw in Chapter 4, bimolecular nucleophilic attack at silicon is less stereospecific and can have linear and adjacent transition

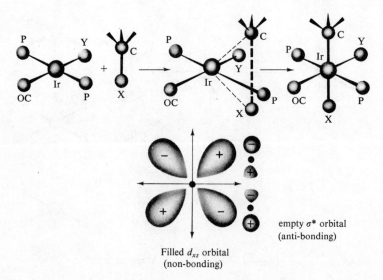

Fig. 10 – 3 Proposed three-centre transition state and orbital diagram for *trans* oxidative addition

states depending upon the electronegativities of the entering and leaving group. Unfortunately, it is not possible to compare the two directly since H_3SiCl (and H_2SiCl_2 and $HSiCl_3$) have the tendency to add across the Si—H rather than the Si—Cl bond. These reactions most certainly go by way of a three-centre transition state

and it is not improbable that the ionizing properties of the solvent (or lack of them) may play a part of considerable importance in determining the course of the reaction. The observation of large negative entropies of activation in the reaction

$$[Ir(diphos)_2]^+ + R_3SiH \longrightarrow cis\text{-}[Ir(diphos)_2SiR_3(H)]^+$$
$$(diphos = Ph_2P\ CH_2CH_2PPh_2)$$

makes one question the validity of the relationship between highly polar transition states for oxidative-addition and the large negative entropies of activation.

The kinetics and stereochemistry of oxidative addition, therefore, indicate that there is a whole spectrum of behaviour between the addition-rearrangement mechanism on the one hand, requiring coordination unsaturation, cis-addition, and non-polar three-centre transition states with retention of configuration at X and Y, and the nucleophilic substitution on the other with a polar transition state that is linear and leads to inversion when X = C, that allows trans as well as cis addition, and which can operate with coordinatively saturated reagents. A great deal of work is now in progress in this area of chemistry in an attempt to work out the rules governing the observed behaviour and a great deal more will be needed before the apparent exceptions and anomalies to any generalized theory have been ironed out.

10-2-3 One-electron oxidative addition

All of the examples discussed until now have related to a two-electron redox process. However, two important groups of reactions involve a one-electron oxidation of the reductant. They are the Cr(II) → Cr(III) (aquo or ammine media) and Co(II) → Co(III) (cyanide media) reactions and can be formulated as

$$2[Cr(H_2O)_6]^{2+} + X-Y \longrightarrow Cr(H_2O)_5X^{2+} + Cr(H_2O)_5Y^{2+}$$
$$2[Co(CN)_5]^{3-} + X-Y \longrightarrow Co(CN)_5X^{3-} + Co(CN)_5Y^{3-}$$

These reactions are closely analogous to the straightforward non-complementary redox reactions involving an inner-sphere redox mechanism and can be so formulated, as for example

$$Co(CN)_5^{3-} + CH_3I \longrightarrow Co(CN)_5I^{3+} + \cdot CH_3$$
$$Co(CN)_5^{3-} + \cdot CH_3 \longrightarrow Co(CN)_5CH_3 \qquad \text{fast}$$

The evidence for the 'free radical' intermediate is strong and alternative fates can await the radical, for example,

$$Co(CN)_5^{3-} + CH_3-\underset{\underset{I}{|}}{CH}-CH_3 \longrightarrow Co(CN)_5I^{3-} + CH_3-\overset{\bullet}{C}H-CH_3$$

$$Co(CN)_5^{3-} + CH_3-\overset{\bullet}{C}H-CH_3 \longrightarrow Co(CN)_5H^{3-} + CH_2=CH-CH_3$$

Side reactions of this sort, together with more equivocal conclusions based on relative reactivities and substituent effects, have fully confirmed the one-electron, 'free radical' mechanism.

10-3 Reductive elimination

This is the reverse reaction of oxidative addition and therefore ought to follow the reverse path in its mechanism. In principle we might again expect two limiting mechanisms, the equivalent of addition redistribution being a unimolecular process in which two ligands join to form a molecule which is then expelled:

$$\begin{array}{c} X \diagdown \underset{PR_3}{\overset{|}{}} \diagup H \\ Ir^{III} \\ OC \diagup \underset{PR_3}{\overset{|}{}} \diagdown H \end{array} \longrightarrow OC-\underset{\underset{PR_3}{|}}{\overset{\overset{PR_3}{|}}{Ir^{I}}}-X + H_2$$

and the equivalent of the nucleophilic attack being a bimolecular attack upon a coordinated ligand and the consequent expulsion of the complex in the lower oxidation state:

$$Cl-\underset{\underset{As}{\diagdown}\underset{As}{\diagup}}{\overset{\overset{As}{\diagup}\overset{As^{2+}}{\diagdown}}{Pt^{IV}}}-Cl \; + SCN^- \longrightarrow NCSCl + \underset{\underset{As}{\diagdown}\underset{As}{\diagup}}{\overset{\overset{As}{\diagup}\overset{As^{2+}}{\diagdown}}{Pt^{II}}} + Cl^-$$

or $PtCl_6^{2-} + I^- \longrightarrow PtCl_4^{2-} + ICl + Cl^-$

Although the consequences of these reductive eliminations are well recognized they have not been investigated kinetically to any large extent apart from cases where the oxidative-addition has been examined under reversible conditions. Any extensive mechanistic discussion is therefore premature, but so far it seems likely that the predictable paths are being followed.

A number of hydride complexes which contain more than one hydride ligand can undergo a reductive elimination in which a hydrogen molecule is expelled as a result of entry by a single ligand. Thus

$$OsH_2Cl_2(PEt_3Ph)_3 + L \longrightarrow OsCl_2(L)(PEt_2Ph)_3 + H_2$$
d^4Os(IV) Seven-coordinate d^6Os(II) Six-coordinate
(L = CO or PMePh$_2$)

Reversible reactions involving nitrogen and hydrogen have been extensively studied in the economically fruitless (so far) search for effective nitrogen fixation agents:

$$CoH_3(PPh_3)_3 + N_2 \rightleftharpoons Co(N_2)H(PPh_3)_3 + H_2$$
d^6Co(III) Six-coordinate d^8Co(I) Five-coordinate

These fall in the same category. The actual mechanistic role of L in promoting this reaction is unclear and it would be of interest to note whether this requires an attack at the metal before dissociation of H_2.

10-4 Redox substitution

In this section we shall consider those processes in which one or more ligands in the coordination shell have been replaced by others coming from outside under circumstances involving a temporary change in the oxidation state of the reaction centre. There are a number of ways in which this can take place of which the following have been documented.

10-4-1 Temporary change in the oxidation state to give a complex that is much more substitutionally labile than the original substrate

An example of this was found many years ago by Taube and Rich who studied the reaction:

$$AuCl_4^- + Cl^{-*} \rightleftharpoons AuCl_3Cl^{*-} + Cl^-$$

and showed that special care was needed in purifying the reagents and solvents before reproducible results were obtained. The catalysed path, for such it was causing the irreproducibility, was shown to be a one-electron reduction to labile Au(II) and a chain reaction was postulated as for example in the Fe^{2+} assisted exchange:

$Au^{III}Cl_4^- + Fe^{2+} \longrightarrow Au^{II}Cl_4^= + Fe^{3+}$ slow initiation
$Au^{II}Cl_4^= + 4Cl^{*-} \longrightarrow Au^{II}Cl_4^{*2-} + 4Cl^-$ ⎫
$Au^{II}Cl_4^{*2-} + Au^{III}Cl_4^- \longrightarrow Au^{III}Cl_4^{*-} + Au^{II}Cl_4^{2-}$ ⎬ fast propogation
$2Au^{II}Cl_4^{2-} \longrightarrow Au^{III}Cl_4^- + Au^{I}Cl_2^- + 2Cl^-$ slow termination

In this particular case the chain is long and for each ferrous ion oxidized some 10^4 acts of exchange were observed. Reactions of this sort are not particularly well documented because the amount of work required to sort out such a system or even to achieve reproducibility is quite considerable. One suspects that many studies where the need for considerable care in the purification of the reagents and solvents has been necessary to achieve reproducibility have been associated with catalytic processes of this sort. Photochemically induced substitution reactions of transition metal complexes often follow a path where the photochemically excited state is substitutionally labile and lasts long enough to replace one or more of its ligands. The photocatalysis of the $PtCl_6^{2-}$—Cl^{*-} exchange is thought to go by chain reaction involving Pt(III) species which are isoelectronic with those of Au(II) reported above.

10-4-2 Substitution as a result of a bridge transfer redox process

To some extent this is related to the previous type of reaction in so far as at least one coordination site on the redox catalyst is substitutionally labile. A very wide range of examples exists in this category, some of which have been mentioned in previous chapters. The Co(II) catalysed displacement of ammonia from $Co(NH_3)_5Cl^{2+}$ by CN^- results in the formation of $Co(CN)_5Cl^{3-}$ even though Cl^- is by far the most labile ligand in the original substrate. In this case, the Co(II) state is substitutionally labile and gives the very strong reducing agent $Co(CN)_5^{3-}$ which reduces the substrate to form the product

$$Co^{2+}aq + 5CN^- \longrightarrow Co(CN)_5^{3-} \quad \text{fast}$$
$$Co(CN)_5^{3-} + Cl-Co(NH_3)_5^{2+} \longrightarrow Co(CN)_5Cl^{3-} + Co_{aq}^{2+} + 5NH_3$$

Another example is the Cr(II) catalysed aquation of one of the chlorides in $CrCl_2(H_2O)_4^+$. This reaction follows a simple second-order rate law (together with the spontaneous first-order process), rate = $k[CrCl_2^+][Cr^{2+}]$, and is simply the standard single bridge inner-sphere redox process

$$(H_2O)_4ClCr^{III}Cl^+ + Cr_{aq}^{2+} \xrightarrow{k} [(H_2O)_4ClCr^{II}]^+ + ClCr^{III}(H_2O)_5^{2+}$$
$$[(H_2O)_4ClCr^{II}]^+ \xrightarrow{fast} Cr_{aq}^{2+} + Cl^-$$

and the measured rate constant for the release of chloride is indeed that for the redox process. It should be pointed out that this type of path cannot catalyse the loss of the remaining chloride since

$$(H_2O)_5Cr^{III}Cl^{2+} + Cr_{aq}^{2+} \longrightarrow Cr_{aq}^{2+} + ClCr^{III}(H_2O)_5^{2+}$$

does not lead to net chemical change. The Cr^{2+} catalysed release of chloride from $CrCl(H_2O)_5^{2+}$ is very slow in comparison and records those processes where chlorine does not constitute the bridge, that is,

$$\text{Cl}(H_2O)_4\text{Cr}-O\begin{matrix}H^{2+}\\ \diagup \\ \diagdown \\ H\end{matrix} + \text{Cr}^{2+}_{aq} \longrightarrow \text{Cl}^- + \text{Cr}^{2+}_{aq} + \text{Cr}(H_2O)^{3+}_6$$

This is very strongly base catalysed since —OH makes a much more effective bridge than —OH$_2$.

The Pt(II) catalysed substitutions of Pt(IV) complexes follow a similar pattern except that only two of the six octahedral positions are substitutionally labile in the product. Thus the reaction

$$\text{Pt en}_2\text{Cl}_2^{2+} + \text{NO}_2^- \longrightarrow \text{Pt en}_2\text{NO}_2\text{Cl}^{2+} + \text{Cl}^-$$

is catalysed by Pt en$_2^{2+}$ by way of the inner sphere redox process.

$$\text{Pt en}_2^{2+} + \text{NO}_2^- \rightleftharpoons \text{Pt en}_2\text{NO}_2^+ \quad \text{fast}$$
$$\text{NO}_2\text{en}_2\text{Pt}^+ + \text{Cl}-\text{Pt en}_2\text{Cl}^{2+} \longrightarrow \text{NO}_2\text{en}_2\text{PtCl}^{2+} + \text{Pt en}^{2+} + \text{Cl}^-$$

Since NO$_2^-$ is a far less effective bridge than Cl in this reaction, the second chlorine cannot easily be replaced by this mechanism

$$\text{NO}_2\text{en}_2\text{Pt}^+ + \text{Cl}-\text{Pt en}_2\text{NO}_2^{2+} \longrightarrow \text{NO}_2 \text{ en}_2\text{PtCl}^{2+}$$
$$+ \text{Pt en}_2^{2+} + \text{NO}_2^-$$

that is, no net change.

10-4-3 Substitution by reversible oxidative addition

In discussing simple substitution it was pointed out that the replacement of one ligand by another involved a temporary change in coordination number at the reaction centre. Since oxidative addition leads to an increase of coordination number and reductive elimination leads to a decrease, a combination of the two, provided that the molecule XY that adds is not identical to XY′ that is eliminated, will lead to a net substitution process in which Y′ is replaced by Y. Obviously the path chosen will depend upon whether oxidation or reduction is more likely.

In the case of d^6 Pt(IV), where simple substitution is not particularly common, there is no chance of an oxidation but a very strong reasonable chance of a reduction. Thus, it has been shown that the reaction between PtCl$_6^{2-}$ and iodide ion is characterized by two intermediates, PtCl$_4$I$_2^{2-}$ and PtCl$_2$I$_4^{2-}$, and that the odd-numbered intermediates are not observed. On the other hand PtI$_6^{2-}$ + Br$^-$ gives PtI$_5$Br^{2-}. All of these observations and the simple second-order kinetics can be explained in terms of a sequence of reversible reductive eliminations, thus

$$\text{PtCl}_6^{2-} + \text{I}^- \longrightarrow \text{PtCl}_4^{2-} + \text{ICl}$$
$$\text{ICl} + \text{I}^- \rightleftharpoons \text{I}_2 + \text{Cl}^- \quad \text{fast—equilibrium lies to the right}$$
$$\text{PtCl}_4^{2-} + \text{I}_2 \longrightarrow \text{PtCl}_4\text{I}_2^{2-} \quad \text{fast}$$

(Substitution of I^- in $PtCl_4^=$ is also possible and could give rise to the odd-number compositions. This would produce a reaction effectively in category (i) that is, more facile substitution in a labile intermediate with a different coordination number.) On the other hand, the odd intermediates are observed when a less electronegative halide is replaced by a more electronegative one.

$$PtI_6^{2-} + Br^- \longrightarrow PtI_4^{2-} + IBr$$

$$IBr + Br^- \rightleftharpoons I^- + Br_2 \qquad \text{fast—but equilibrium lies to the left}$$

$$PtI_4^{2-} + IBr \longrightarrow PtI_5Br^{2-} \qquad \text{fast}$$

In certain substitution reactions of Pt^{II} the first stage is an oxidative addition. Thus the exchange of hydrogen by deuterium in trans-$[PtH \cdot Cl(PEt_3)_2]$ is acid catalysed and extremely sensitive to the nature of the anion. It takes place readily in non-polar solvents and kinetics indicate a mechanism of the type

$$\textit{trans-}Pt^{II}(PEt_3)_2HCl + DCl \rightleftharpoons Pt^{IV}(PEt_3)_2HDCl_2$$
$$Pt(PEt_3)_2HDCl_2 \rightleftharpoons \textit{trans-}[Pt(PEt_3)_2DCl] + HCl$$

A similar mechanism is also expected in many reactions involving the breaking of metal–carbon bonds, thus, trans-$[Pt(PEt_3)_2(C_6H_5)_2]$ reacts with HCl in methanol to give trans-$[Pt(PEt_3)_2(C_6H_5)Cl] + C_6H_6$. The rate law is rate = $k[\text{complex}][H^+]$ and is independent of the concentration of chloride. Nucleophilic displacement is ruled out and it has been suggested that the rate-determining step is addition of $CH_3OH_2^+$:

$$Pt(PEt_3)_2(C_6H_5)_2 + CH_3OH_2^+ \xrightarrow{\text{slow}} Pt(PEt_3)_2(C_6H_5)_2(H)(CH_3OH)^+$$
$$Pt(PEt_3)_2(C_6H_5)_2(H)(CH_3OH)^+ \xrightarrow{\text{fast}} Pt(PEt_3)_2(C_6H_5)(CH_3OH)^+ + C_6H_6$$
$$Pt(PEt_3)_2(C_6H_5)(CH_3OH)^+ + Cl^- \xrightarrow{\text{fast}} Pt(PEt_3)_2(C_6H_5)Cl + CH_3OH$$

This mechanism was preferred to a direct electrophilic attack on the carbon atom.

The same mechanism of reversible oxidative addition can be assigned to stages in certain metal catalysed reactions of the ligand, or potential ligands. Some of these will be discussed in the next chapter.

Problems

10-1 When purified trans-$[Co\ en_2Cl_2]Cl$ is reacted with excess aqueous KCN the initial product of the reaction is trans-$[Co\ en_2OH\ Cl]Cl$ but if the crude complex (generally contaminated with Co(II)) is used, a major product is $[Co(CN)_5Cl]^{3-}$. Explain.

10-2 In the reaction between $PtBr_6^{2-}$ and excess iodide, the product is PtI_6^{2-}, and trans-$[PtBr_4I_2]^{2-}$ and trans-$[PtBr_2I_4]^{2-}$ can be identified as intermediates. The reverse reaction, $PtI_6^{2-} + Br^-$, does not go through the same intermediates. Suggest a plausible mechanistic explanation. [See E. J. Bounsall, D. J. Hewkin, D. Hopgood, and A. J. Pöe, *Inorg. Chim. Acta*, 1967, **1**, 281.]

10-3 In the presence of $[Pt\ en_2]^{2+}$, the reaction between trans-$[Pt\ en_2Cl_2]^{2+}$ and NO_2^- obeys the rate law, rate = $k[Pt(IV)][Pt(II)][NO_2^-]$. Write down a suitable mechanism for the reaction. Explain why the product of the reaction is trans-$[Pt\ en_2NO_2Cl]^{2+}$ and why the dinitro complex cannot be formed by this mechanism. [See H. R. Ellison, F. Basolo, and R. G. Pearson, *J. Am. Chem. Soc.*, 1961, **83**, 3943.]

11 Catalysis and conclusions

11-1 Catalysis

Until now the contents and emphasis of this book have been academic, in the sense that we have confined our discussion to basic questions such as 'How does the chemical change occur?', 'What factors control its behaviour?', and so on, and we have attempted to understand and classify the answers to these questions in terms of changes in coordination number, coordination geometry, and oxidation state. At no stage have we really considered the relationship between this information and the so-called problems of relevance, such as how to use inorganic complexes to catalyse reactions between small molecules in order to synthesize specific compounds on a large scale or how to polymerize propylene in a stereospecific way (to know how to depolymerize such polymers, whether stereospecifically or not, may be of even greater value since it might provide a means of cutting down the littering of the countryside and the seas by discarded polythene bags and containers). It would be foolish to believe that, armed with the knowledge of the fundamentals and the patterns of behaviour, we can now sally forth from our academic ivory towers and apply our understanding to the alleviation of the world's problems (hoping on the way to benefit from any patents that might accrue). If we look into the real world we find that we come late on the scene and the best we can do is to explain the mechanisms of reactions, many of which have been developed by empirical methods (or even by accident), but all of which have been successfully producing the goods for many years. It may be possible to suggest improvements or even devise new systems along the same lines but, all in all, the applications come, as ever, before the understanding.

The purpose of this book, unashamedly academic, has now been achieved but the temptation remains to show how combinations or sequences of the fairly simple reactions that have been discussed can lead to processes of considerable importance to us all. We can, of course, consider new types of processes that bring cash and prosperity to the chemical industry (assuming of course that they are capable from time to time of making correct decisions of management and planning). We may gain tremendous benefit from these products (although, on occasions, quite the reverse may operate). We should also bear in mind that many biological systems intimately involve metal ions and the active centre may be looked upon as a metal complex. The term 'reactions of coordinated ligands' may be applied to systems that play an important or even a dominating part in processes such as oxygen transfer (haemoglobin involves an iron complex of a macrocyclic quadridentate nitrogen donor ligand), photosynthesis (chlorophyll is a magnesium complex of another macrocyclic quadridentate nitrogen donor), and many enzymatic processes.

In order to give some idea of what is currently of interest, four non-biological topics have been chosen for a brief survey.

11-2 Fixation of atmospheric nitrogen: a problem yet to be solved

Eighty per cent of the gaseous environment consists of N_2, which would make an admirable raw material for the synthesis of nitrogen-containing materials. Certain organisms have the knack of fixing atmospheric nitrogen at standard conditions of temperature and pressure without any difficulty whatsoever.

The simplest fixation, namely with oxygen or hydrogen, leads to products which are thermodynamically stable but, in all cases, somewhere along the line the $N\equiv N$ bond ($= 225\cdot 8$ kcal.mol^{-1}) must be broken. This will, of course, be compensated in the final accounting by the new bonds that have been formed in the stable products, but when we consider the actual rates of formation the activation energies of the simple processes will be high.

In the synthesis of ammonia from hydrogen and nitrogen thermodynamics are on our side in the sense that for

$$N_2 + 3H_2 \rightleftharpoons 2NH_3$$

the free energy change ($\Delta G^0_{298} = -3\cdot 9$ kcal.mol^{-1}) is favourable and the equilibrium lies well to the right at room temperature. The rate at which this equilibrium is achieved is infinitely slow because there is no way of compensating for the fission of the $N\equiv N$ bond [the H—H bond (103 kcal.mol^{-1}) presents fewer problems]. While raising the temperature may increase the rate, it also shifts the position of equilibrium away from the favourable right-hand side and high pressures are required to shift it back again. The Haber process uses inefficient catalysts but they work after a fashion although temperatures in the region of 400–550°, and consequently pressures of 100–1000 atmospheres, are necessary. The utilitarian aim of current research in nitrogen fixation would be to catalyse the reaction so that conversion would be efficient at room temperature and atmospheric pressure. The extension of the process from the simple hydrogen-nitrogen reaction, to reactions involving nitrogen, unsaturated organic molecules, hydrogen, and CO to give a wide range of nitrogen-containing organic materials would be relatively simple, once the first stage has been overcome. As we saw in the last chapter, the activation of hydrogen, whereby the H—H bond is broken in an oxidative addition, is well established. Thus the reaction

$$Co^I H(PPh_3)_3 + H_2 \longrightarrow Co^{III} H_3 (PPh_3)_3$$

splits the H—H bond but, although the process $CoH_3(PPh_3)_3 + N_2 \rightarrow CoHN_2(PPh_3)_3 + H_2$ takes place quite readily, the nitrogen remains coordinated as dinitrogen (N_2) and the required activation is not achieved. A wide range of dinitrogen complexes (N_2 is isoelectronic with CO) have now been described. Of course, $H_2 \rightarrow 2H^-$ is a two-electron reduction (which is quite normal), but $N_2 \rightarrow 2N^{3-}$ is a six-electron reduction and must either involve a multistage process or else a catalyst that can span such a range of oxidation state change. A metal which forms dinitrogen complexes in low oxidation states and nitride complexes in high oxidation states is an obvious candidate for close examination and much work is being done with the complexes of the Fe, Ru, Os triad.

Success, such as it is, has been achieved with the earliest transition elements and the wholly empirical approach seems to have been adopted, at least at the beginning. $Ti(\pi-C_5H_5)_2Cl_2 + C_2H_5MgBr$ in ether will absorb nitrogen at room temperature with breakage of the N≡N bond but the ammonia is not easily released and the investigators seem to be happy if they can achieve a yield of one ammonia per metal atom present. A whole range of early transitional metal complexes plus organo-metallic reducing agents have been tried in this way and, although the first part of the task has been achieved after a fashion, and the nitrogen—nitrogen bond has been broken at room temperature, the second part of the problem, namely to convert the 'activated' nitrogen into a suitable compound while regenerating the catalyst, has yet to be solved.

Examination of the protein that has been associated with azobacter Vinelandii nitrogenase shows molybdenum, iron, and sulphur ligands are present and essential for the nitrogen fixation. Work is now reported on the nitrogen fixation properties of empirical brews, for example, sodium molybdate plus 1-thioglycerol and ferrous sulphate together with a pinch of sodium borohydride. That this functions at all is somewhat surprising and rather akin to constructing a chronometer from iron, copper, gold, silica, etc., because components extracted from a broken up watch analysed for these materials. We are still a long way from a process in which nitrogen and hydrogen are passed in at one end of the catalytic system and ammonia emerges at the other.

From the economic point of view, any low temperature-low pressure catalytic system for the formation of ammonia will have to be very cheap indeed to compete with the Haber process where the major cost is capitalization of high pressure plant. Faced with the problem of a new and simple production method the present ammonia producers would not take kindly to the prospects of writing off their highly expensive high-pressure plant, and the cost margin would have to be enormous to persuade them to do this. It is much more likely that any new successful fixation method would be best applied to the direct synthesis of organo-nitrogen compounds.

11-3 Polymerization of alkenes and alkynes

Unlike N_2, olefins are chemically reactive and the problem is not to use catalysis to induce reactivity but rather to use it to channel the reaction in a very specific direction. Polymerization is favoured thermodynamically, for example:

$$nCH_3CH_2=CH_2 \longrightarrow (-\underset{\underset{CH_3}{|}}{CH}-CH_2-)_n$$

$\Delta H = -20$ kcal mol of propylene polymerized

and can take place by ionic or free radical mechanisms.

(i) Anionic polymerization:

$$X^- + CH_2=CH_2 \longrightarrow X-CH_2-CH_2^-$$
$$XCH_2CH_2^- + CH_2=CH_2 \longrightarrow X-CH_2CH_2CH_2CH_2^- \quad \text{and so on}$$

(Rare because few nucleophiles are attracted by the normal olefinic bond.)

(ii) Cationic polymerization:

$$H^+ + CH_2=CH_2 \longrightarrow CH_3CH_2^+$$
$$CH_3CH_2^+ + CH_2=CH_2 \longrightarrow CH_3CH_2CH_2CH_2^+ \quad \text{and so on}$$

(iii) Free radical polymerization:

ROOR $\longrightarrow$ 2RO·   initiation
RO· + $CH_2=CH_2 \longrightarrow ROCH_2CH_2^\bullet$
$ROCH_2CH_2^\bullet + CH_2=CH_2 \longrightarrow ROCH_2CH_2CH_2CH_2^\bullet$ and so on

Ethylene is a bad example to quote, because it is very reluctant to undergo these reactions and the olefin requires to be activated by the appropriate substituents.

The initial observations of Ziegler that aluminium trialkyls, which were potential Lewis acid (electron pair acceptor) catalysts, reacted with ethylene to produce oligomers (polymers in which only a few monomers are combined) were followed by the development of the so-called Ziegler-Natta catalysts in which adding a transition metal halide to a main group organo-metallic compound led to heterogeneous catalysts which could not only polymerize the generally unreactive ethylene but do so in a way that led to very regular, high molecular weight, almost crystalline high density material. While people were still trying to work out the mechanism or even to identify the actual catalytic species, the process was producing polythene and polypropylene, and so on, on a very large scale.

The general features of the mechanism are quite clear. The new olefin monomer is inserted into the growing polymer chain at its point of attachment to the metal site in the catalyst. The catalyst requires a labile site which can be occupied by the entering olefin and it has been strongly suggested that the initial bonding is similar to that encountered in the stable olefin complexes of the later transition elements.

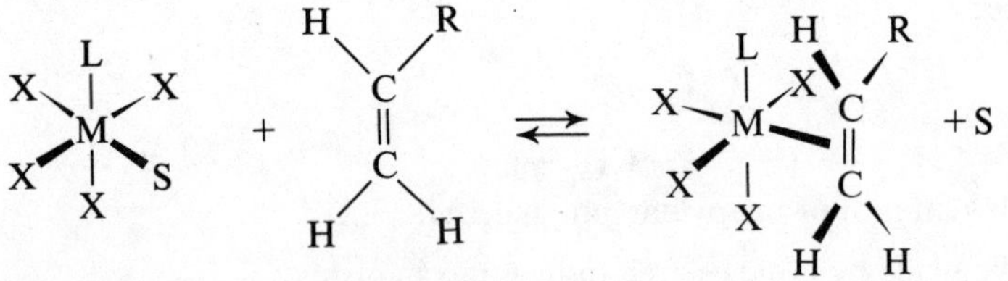

The catalytically important step is the change in bonding from the original π-interaction to an M—C σ-bond which is accompanied by an effective migration of L from M to C

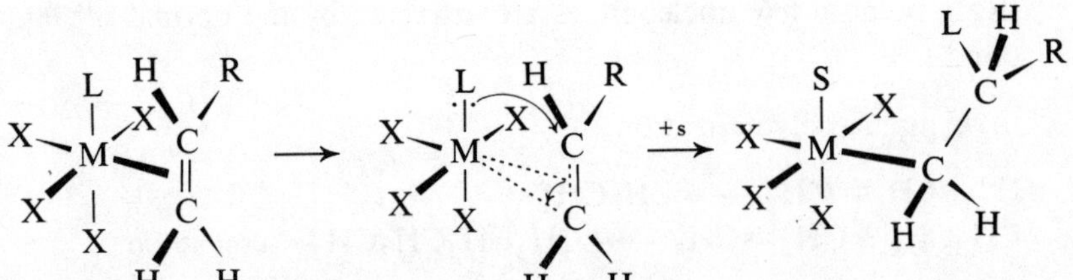

The ligand migration (or insertion) reaction has not been studied specifically in this particular reaction but has received extensive attention elsewhere since the ligand migrating can be any one of a very large number of functional groups and this type of reaction far more extensive in its application than olefin polymerization. The distinction between ligand migration and insertion is that, in the former case, as shown above, the growing chain moves to the *cis* position that was occupied by the entering monomer, while in the insertion mechanism it stays where it is. By using an appropriately signposted complex the distinction can be clearly made.

Polymerization of alkenes and alkynes

This has been done, for example, in the reaction

$$trans\text{-}[Mo(CO)_4(CH_3CO)(PPh_3)] \underset{}{\overset{heat}{\rightleftharpoons}} cis\text{-}[Mo(CO)_4(CH_3)(PPh_3)] + CO$$

where the triphenyl phosphine acts as the signpost:

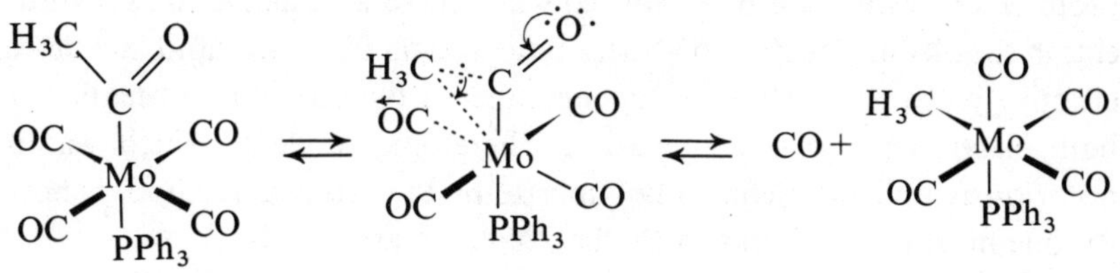

trans acyl ⇌ ⇌ CO + *cis* alkyl

It has also been shown that the act of ligand migration requires simultaneous entry of a ligand or a solvent molecule into the coordination shell of the metal ion to take up the vacancy that would otherwise be created. The normal path for chain breaking is the reverse of the ligand migration/insertion process, but with C—H rather than C—C bond breaking:

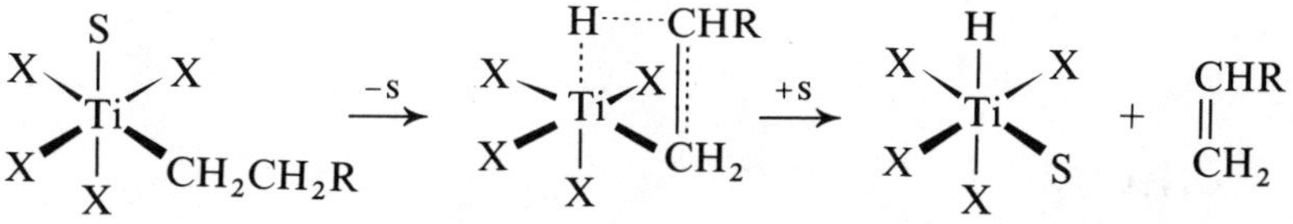

This also leaves the catalyst in a form suitable to start a new chain:

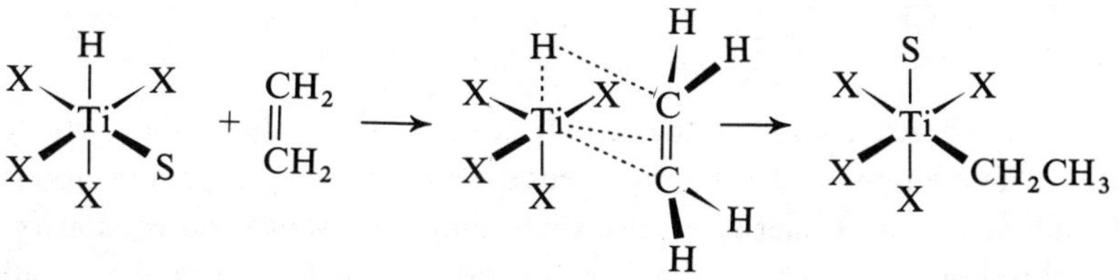

The average chain length will depend upon the relative chances of hydride formation (which is the chain terminating step) and olefin insertion/alkyl migration (which is the chain lengthening process). The alkyl polymer can be released from the catalyst by reaction with a protolytic reagent but the catalyst will require further regeneration as a consequence.

174 Catalysis and conclusions   Ch. 11

The effectiveness of a catalyst is very sensitive to the nature of the ligands around the reaction centre (in so far as the actual catalytic species has been identified for many heterogeneous Ziegler-Natta catalysts). It is necessary that the metal-olefin bond is strong enough to attract the olefin to the reaction centre, and yet the metal—carbon σ-bond must be weak enough to allow facile ligand migration. It is of interest to point out that catalysis of cationic polymerization of olefins using simple protonic or Lewis acids only worked when there were activating substituents in the olefin. The Ziegler-Natta catalysts, on the other hand, are more effective when the olefins are unencumbered by many substituents, for both electronic and steric reasons. Therefore, ethylene, which was a notoriously difficult olefin to polymerize by the classical method, is now the one most readily handled by the Ziegler-Natta catalysts.

When propylene, $CH_3CH=CH_2$, is polymerized, the choice of orientation of each monomer with respect to the growing chain is quite considerable. Protonic and Lewis acid catalysts generally produce oily or waxy polymeric products of relatively low molecular weight because proton transfer involving the growing carbonium ion leads to extensive chain branching. The Ziegler-Natta catalysts prevent this by eliminating the need for carbonium ion formation. Furthermore, the olefin monomers will all be lined up the same way round since the act of insertion forms a bond between the metal and the least substituted carbon and the growing chain migrates to the other end of the double bond.

$$\begin{array}{c} CH_2 \\ | \\ CH.CH_3 \\ | \\ CH_2 \end{array} \quad \begin{array}{c} H \\ | \\ X\diagdown \;|\; \diagup X \diagup C \diagdown CH_3 \\ Ti \\ \diagup \;|\; \diagdown \;\; \| \\ X \;\;\; X \;\;\; CH_2 \end{array} \longrightarrow \begin{array}{c} S \\ X\diagdown \;|\; \diagup X \\ Ti \\ \diagup \;|\; \diagdown CH_2CH(CH_3)CH_2CH(CH_3)CH_2 \ldots \\ X \;\;\; X \end{array}$$

A further choice comes in the actual configuration about the tertiary carbon. There may be a random orientation of the methyl groups about the polymer chain (**atactic**) or else there may be a structural regularity. The simplest two such regular arrangements are the cases where all tertiary carbons have the same configuration (**isotactic**) or have alternating configurations (**syndiotactic**). In both cases, the regular structure coupled with the absence of chain branching makes a 'crystalline' polymer of relatively high density, hardness, and melting or softening point (Fig. 11-1). The ability of certain heterogeneous Ziegler-Natta catalysts to produce isotactic polypropylene has been ascribed to the existence

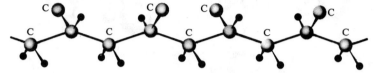

(a) Isotactic—all chain carbons have the same configuration

(b) Syndiotactic—regular alternation of configuration

(c) Atactic—no regularity of configuration

Fig. 11 – 1 Linear polymers of propene

of site vacancies in the surface adjacent to the metal catalytic centre (a missing chloride ion, for example). These allow the methyl group to 'snuggle' into the site and ensure that the propylene molecule is always oriented the same way when it is inserted into the growing chain. The mechanism must also require the growing chain to move back to its original position on the surface before the next monomer is incorporated. In general, the preparation of isotactic polymers is very difficult except with heterogeneous catalysts and soluble homogeneous catalysts generally produce atactic polymers. However, it has been possible to produce homogeneous catalysts that will generate syndiotactic polymers at low temperatures. A catalyst consisting of VCl_4 in hexane, to which $AlEt_2Cl$ has been added is quite effective, especially when a Lewis base, such as anisole is also present and the reaction carried out below $-80°$. The mechanism suggested is precisely that which has been already described above, the only modification being the presence of ligand Y to orient (by steric effects) the propylene molecule and a low enough temperature to restrict rotation about the V—C bond of the growing polymer (Fig. 11-2). It is the alternation of the position of the growing chain between the two sites relative to ligand Y that causes the alternation of configuration.

Another type of oligomerization involving alkynes was first developed by Reppe using as catalysts nickel(II) complexes with a range of bidentate ligands such as acetylacetone, salicylaldehyde, and so on, the requirement being that the nickel remained high-spin octahedral (or tetrahedral) and

176 Catalysis and conclusions Ch. 11

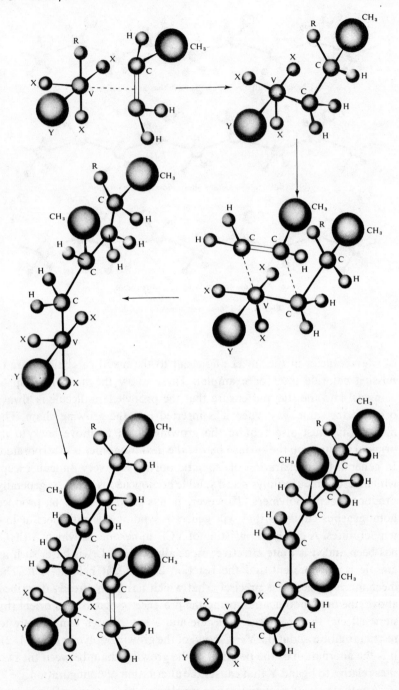

Fig. 11 – 2 Formation of a syndiotactic polymer as a result of ligand migration. The bulk of Y controls the orientation of the monomer

that substitution was rapid (a reasonable requirement for a catalyst). The normal reaction was the tetramerization of acetylene to give cyclooctatetraene (which might have been a disappointment if the object of the exercise was to catalyse the formation of benzene and provide a catalytic pathway for the synthesis of aromatics). It is thought that four acetylene molecules occupy four of the six coordination sites about the metal and that the bonding rearranges simultaneously to allow cyclization (Fig. 11-3). This is in preference to a stepwise oligomerization which would yield a wider variety of products. The reaction can be modified (and the original purpose vindicated) by adding a competing ligand. Triphenylphosphine in 1:1 molar ratio with the catalyst inhibits the formation of cyclooctatetraene and benzene is formed instead. This is fully in keeping with the idea that only three coordination sites are available. Bidentate inhibitors, such as $Ph_2PCH_2CH_2PPh_2$, phen, bipy, ruin the catalyst completely and no further polymerization takes place and here it is suggested that only two *trans* sites are left and the acetylene molecules occupying them cannot interact.

This type of oligomerization is of considerable importance and suitable modification of the catalyst and the alkene can lead to the synthesis of a wide range of conjugated products. Cyclodimerization and trimerization of diolefins such as butadiene to give cyclooctadiene and cyclododecatriene can be catalysed by nickel(0) complexes and the interference of the simple path by blocking coordination positions by suitable ligands leads to a wide variety of products. In most of the polymerization work, the devising of new catalysts savours more of craftsmanship than science. The catalysts are none the worse for their parentage and do their jobs every bit as well as if they were well born, aristocratic, purposefully synthesized molecules of known composition and structure.

11-4 Hydrogenation of alkenes

The addition of hydrogen to olefins is of considerable importance in a variety of processes, but it requires catalysis and quite often high pressures of hydrogen. In recent years homogeneous catalysts have been developed that are clean, often selective, and by keeping their molecular integrity allow a study to be made of the detailed mechanism. A typical catalyst, now marketed for the purpose, is $Rh(PPh_3)_3Cl$ which as a labile Rh(I) species, is very sensitive to oxidative addition. The mechanism of hydrogenation catalysis by this species demonstrates the consequences of the combination of the various primary steps that have been discussed in previous chapters. A study of the kinetics indicates that the rate of hydrogenation is retarded by excess triphenyl phosphine and a pre-equilibrium solvolysis is indicated:

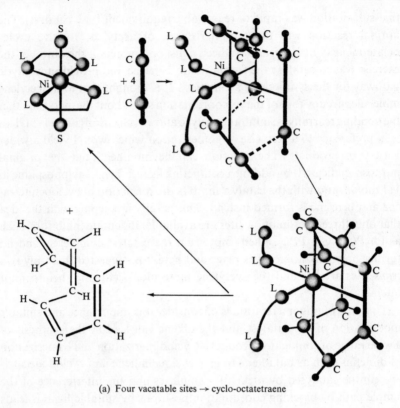

(a) Four vacatable sites → cyclo-octatetraene

Fig. 11 – 3 Tetramerization and trimerization of acetylene by Ni(II) catalysts

$$RhCl(PPh_3)_3 + S \rightleftharpoons RhCl(PPh_3)_2S + PPh_3$$

Early measurements of the molecular weight of this catalyst in benzene suggested that this equilibrium was very much over to the right (which would conflict with the kinetic observations) but recently it has been shown that the anomalous molecular weight was due to the presence of oxygen and the reaction

$$RhCl(PPh_3)_3 + O_2 \longrightarrow RhCl(PPh_3)_2O_2 + PPh_3$$

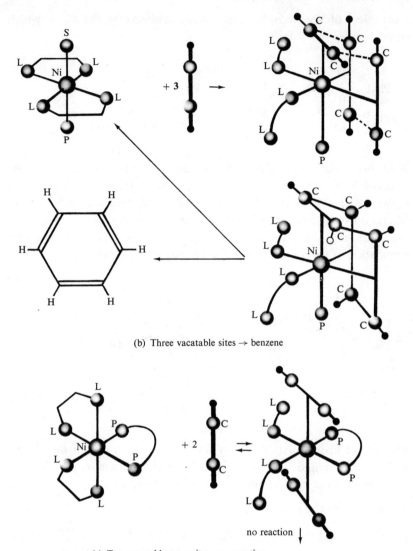

(b) Three vacatable sites → benzene

(c) Two vacatable *trans* sites—no reaction

The next stage in the reaction is the oxidative addition of hydrogen to form the *cis* dihydrido Rh(III) species:

$$\begin{array}{c}\text{Ph}_3\text{P}\diagdown\diagup\text{S}\\ \text{Rh}^{\text{I}}\\ \text{Ph}_3\text{P}\diagup\diagdown\text{Cl}\end{array} + \text{H}_2 \longrightarrow \begin{array}{c}\quad\text{H}\\ \text{Ph}_3\text{P}\diagdown\ |\ \diagup\text{H}\\ \text{Rh}^{\text{III}}\\ \text{Ph}_3\text{P}\diagup\ |\ \diagdown\text{S}\\ \quad\text{Cl}\end{array}$$

The solvent molecule, labile already but further labilized by the strong *trans* effect of the phosphine, is readily replaced by the olefin which enters bonded in a π-fashion:

$$\begin{array}{c} Ph_3P\diagdown \overset{H}{\underset{Cl}{\overset{|}{Rh}}}\diagup \overset{H}{\underset{S}{}} + \overset{CHR}{\underset{CHR'}{\|}} \longrightarrow Ph_3P\diagdown \overset{H}{\underset{Cl}{\overset{|}{Rh}}}\diagup \overset{H}{\underset{CHR'}{\overset{CHR}{\|}}} + S \end{array}$$

It has been shown clearly that, in this case, the olefin does enter into the oxidized form of the catalyst and that the side reaction in which olefin replaces S or PPh_3 in the Rh(I) complex does not lead to hydrogenation. Migration of hydride accompanies the π → σ change in the bonding of the olefin and simultaneously or very soon after there is a reductive elimination of the alkane and the regeneration of the catalyst.

$$Ph_3P\diagdown \overset{H\cdots}{\underset{Cl}{\overset{|}{Rh^{III}}}}\diagup \overset{H}{\underset{CHR'}{\overset{CHR}{\|}}} \xrightarrow{s} Ph_3P\diagdown \overset{S}{\underset{Cl}{\overset{|}{Rh^{III}}}}\diagup \overset{H}{\underset{CHR'CH_2R}{}}$$

$$\longrightarrow Ph_3P\diagdown Rh^{I}\diagup \overset{S}{\underset{Cl}{}} + CH_2R'CH_2R$$

The rate of hydrogenation is very sensitive to the nature of the ligand. Triphenylphosphine is reasonably good, and the rate can only be doubled by greatly increasing the basicity of the phosphine, as in

$(p-CH_3O\cdot C_6H_4)_3P.$

An electron-withdrawing substituent, as in $(p-Cl-C_6H_4)_3P$, reduces the rate considerably (factor of 25 in the hydrogenation of cyclohexane). Ph_3As is less effective than the phosphine and the Ph_3Sb complex is not catalytically active. Basicity arguments are only of limited value and the strong dependence of rate on the non-participating ligands can be explained in terms of facilitation of oxidative addition or labilizing of the Rh—H and Rh—C bonds. However, since one of the phosphine ligands has to leave the complex it is possible to have too much of a good thing, since, if the basicity of the ligand is increased too much, the solvolytic dissociation will become difficult. Steric features may also play a part in this initial solvolysis since the less bulky trialkyl phosphines are found to produce less effective catalysts. The $RhClL_3$ catalyst is effective and fairly unselective in its choice of olefins to hydrogenate but $RhH(CO)(PPh_3)_3$,

Hydrogenation of alkenes

when it works, is specific for terminal olefins only (that is R′ = H). Part of the problem is steric, in that substituents will get in the way of the two *trans* phosphines, and part of the problem is electronic since the CO ligand is in competition with a π-bonding ligand in the *trans* position and only the most avid olefins (that is, terminal olefins) can compete adequately. The mechanism of this catalysis is considerably different from that already mentioned. The first step is the loss of a phosphine,

$$RhH(CO)(PPh_3)_3 \rightleftharpoons \underset{CO}{\overset{Ph_3P}{}} Rh \underset{PPh_3}{\overset{H}{}} + PPh_3$$

and this differs from the previous case in that the two remaining phosphines are *trans*. The second major point of difference is that the olefin attaches to the metal when the latter is in the lower oxidation state and forms a five-coordinate species:

$$\underset{OC}{\overset{Ph_3P}{}} Rh^I \underset{PPh_3}{\overset{H}{}} + \underset{CH_2}{\overset{CHR}{\|}} \longrightarrow \underset{OC}{\overset{Ph_3P}{}} Rh^I \overset{\overset{CHR}{\underset{\|}{CH_2}}}{\underset{PPh_3}{\overset{H}{}}}$$

Hydride migration also takes place in this oxidation state,

$$\underset{OC}{\overset{Ph_3P}{}} Rh \overset{\overset{CHR}{\underset{\|}{CH_2}}}{\underset{PPh_3}{\overset{H}{}}} \rightleftharpoons \underset{OC}{\overset{Ph_3P}{}} Rh \underset{PPh_3}{\overset{CH_2CH_2R}{}}$$

and the oxidative addition of hydrogen is then accompanied by the reductive elimination of the alkane

$$\underset{OC}{\overset{Ph_3P}{}} Rh^I \underset{PPh_3}{\overset{CH_2CH_2R}{}} + H_2 \rightleftharpoons \underset{OC}{\overset{Ph_3P}{}} \underset{PPh_3}{\overset{\overset{H}{|}}{Rh^{III}\underset{|}{}}} \overset{CH_2CH_2R}{\underset{H}{}}$$

$$\rightleftharpoons \underset{OC}{\overset{Ph_3P}{}} Rh^I \underset{PPh_3}{\overset{H}{}} + CH_3CH_2R$$

The reversibility of the earlier part of the process provides a pathway whereby an olefin can isomerize and the relative importance of isomerization and hydrogenation depends on whether or not the reverse σ-alkyl-π-alkene + hydride process is more favoured than the following oxidative addition reductive elimination process.

11-5 Hydroformylation reaction

It was before the war that Roelen discovered a reaction in which an olefin, hydrogen, and carbon monoxide combined together under the catalytic influence of cobalt carbonyl to form aldehydes, and yet people were still speculating as to the nature of the mechanism well into the nineteen-sixties. The general features of the mechanism combine π-bonding of olefins changing to σ-bonding on ligand migration, carbonyl insertion, oxidative addition by hydrogen, and then reductive elimination. The effective low oxidation state catalyst is $HCo(CO)_4$ which if formulated as Co(I) is coordinatively saturated and requires ligand dissociation before it can accept the olefin:

$$HCo(CO)_4 \rightleftarrows HCo(CO)_3 + CO$$

$$HCo(CO)_3 + \underset{CH_2}{\overset{CRH}{\|}} \longrightarrow \underset{CH_2}{\overset{CRH}{\|}} \text{—} \overset{H}{\underset{}{Co(CO)_3}}$$

The now familiar π → σ transformation takes place, accompanied by migration of the hydride from Co to C:

$$\underset{CH_2}{\overset{CRH}{\|}} \text{—} \overset{H}{\underset{}{Co(CO)_3}} \longrightarrow RCH_2CH_2Co(CO)_3$$

π-olefin   σ-alkyl

The migration of the alkyl group from Co to CO (as in the case of the alkyl → acyl interconversions on Mn and Mo) is facilitated by the entry of CO:

$$RCH_2CH_2\underset{\underset{\sigma\text{-alkyl}}{\overset{\cdot\cdot\cdot}{C=O}}}{\overset{}{-}}Co(CO)_2 + CO \longrightarrow RCH_2CH_2\underset{\underset{\sigma\text{-acyl}}{\overset{\|}{O}}}{\overset{}{-}}\overset{}{C}\text{—}Co(CO)_3$$

The hydrogenation, in the last step, liberates the aldehyde and regenerates the catalyst,

$$\text{RCH}_2\text{CH}_2\underset{\underset{\text{O}}{\|}}{\text{C}}-\text{Co(CO)}_3 + \text{H}_2 \longrightarrow \text{HCo(CO)}_3 + \text{RCH}_2\text{CH}_2\text{CHO}$$

The mechanism of this last step could be a reversible oxidative addition of hydrogen but many other mechanisms can equally well be formulated.

Analogous carbonyl complexes of the congeners Rh and Ir are far less successful, partly because of the greater tendency to form metal–metal bonds, but replacement of CO by other ligands, such as phosphines, leads to some very effective Rh(I) hydroformylation catalysts. $RhCl(PPh_3)_3$, which has already been mentioned in its role as a hydrogenation catalyst, is also effective in hydroformylation but although added as such it immediately undergoes ligand substitution and the active catalyst in some other species. A much more recognizable catalyst is $RhH(CO)(PPh_3)_3$, the five-coordinate hydrogenation catalyst mentioned in the last section, whose fate during the reaction can be guessed from the kinetics and by the isolation of stabilized intermediates or their analogues. These Rh(I) catalysts are more effective than the original $Co_2(CO)_8$, which needed fairly high pressures, and can be run at normal or only slightly elevated temperatures. The mechanism that has been put forward for the action of $RhH(CO)(PPh_3)_3$ is as follows:

$$RhH(CO)(PPh_3)_3 + CH_2{=}CHR \xrightarrow[-PPh_3]{(1)}$$

$$Rh(CH_2CH_2R)(CO)(PPh_3)_2 \xrightarrow[-PPh_3]{(2)}$$

$$Rh(CH_2CH_2R_3)(CO)(PPh_3)$$

$$(3)\ -CO \Updownarrow +CO$$

$$Rh(COCH_2CH_2R)(CO)_2(PPh_3)_2$$

$$(4)\ -H_2 \Updownarrow +H_2$$

$$RhH_2(COCH_2CH_2R)(CO)(PPh_3)_2$$

$$(5) \Updownarrow$$

$$RhH(CO)(PPh_3)_2 + \underset{\underset{\text{O}}{\|}}{\text{H}-\text{C}}\text{CH}_2\text{R}$$

Step (1) is probably a typical π-olefin to σ-alkyl conversion with the hydride ligand migrating to the β-carbon, step (2) is an unproductive pathway whereby the catalyst is destroyed, step (3) is a standard alkyl migration-carbonyl insertion resulting from substitution by another CO, step (4) is oxidative addition of hydrogen to oxidize the Rh(I) to Rh(III), and step (5) is the reductive elimination that generates a C—H bond, converts the coordinated acyl group to aldehyde, and regenerates the original catalyst. It is possible that (4) and (5) are synchronous. By using fluoro olefins, which form stable Rh—C bonds or using analogous iridium complexes which also are less effective because of the more stable metal—carbon bonds, Wilkinson has been able to isolate and characterize many analogues of the intermediates postulated in the above reaction scheme and examine their behaviour independently, for example

$Rh(CF_2CF_2H)(CO)(PPh_3)_2$
$Ir(Ph)(CO)(PPh_3)_2$
$Rh(CF_2CF_2H)(CO)_2(PPh_3)_2$
$Ir(R)(CO)_2(PPh_3)_2$ (R = Et, Ph)
$Ir(CO)(COCH_2CH_3)(PPh_3)_2$
$Rh(CO)_2(COR)(PPh_3)_2$ (R = Et, Ph)
$Ir(CO)_2(COR)(PPh_3)_2$ (R = Et, Ph)

11-6 What of the future?

The subject matter of this book has been concerned with the state of the subject as it was or as it is. The competence of an author in presenting this sort of information reflects his ability to keep up with the chemical literature and his understanding of what he finds there. But the subject of inorganic reaction mechanisms is developing all the time and, to judge by trends in the numbers of papers published, accelerating rapidly. To predict the future therefore requires a measure of clairvoyance; to record one's predictions in a book that may still be available in a corner of a dusty bookshelf in a library ten years hence requires a measure of foolhardiness. But a book like this ought to finish with some sort of view as to what lies ahead, however wrong this view might appear to be eventually. It is convenient to separate the academic aspects from the applied aspects, although this might be a direct contradiction of future trends where it seems that research finance is becoming more and more diverted towards 'utilitarian' lines and will remain so for a number of years to come.

It is easy to predict that there will be a great deal of 'handle turning' with more and more information about conforming systems that are reasonably well understood now. This will lead to bigger and better tables of data. More originally minded people will turn their attention to new geometries and coordination numbers (five, seven, and eight) as the compounds and techniques for study become more available and understood. The emphasis of fast reaction studies will move more towards the borderline of coordination and solvation and the formation, structure,

Sec. 11–6   What of the future?   **185**

and geometry of outer-sphere complexes will be a subject of considerable interest. This in turn ought to provide a new approach to the general problem of solvation and it may then be realized that the potassium ion in aqueous solution is a very poor model system to study and that the $Cr(H_2O)_6^{3+}$ ion, with its inert coordination shell, is a much better model upon which to fit the surroundings of solvent and other solute species. The studies of intramolecular stereochemical change will become more and more elaborate and it seems reasonable to believe that the facts may start to catch up with the speculation. The treatments of the 'mathematics' and geometry of intramolecular isomerization is well advanced and what is now needed is an adequate system of signposting within the complex (and means to analyse with accuracy the information that this produces). If it is ever considered that the subject of stereochemistry and stereochemical change in the undergraduate syllabus be extended to include these topics and the problems of stereochemical non-rigidity and fluxional molecules in general, there is going to be a major problem of visualization. The lecturer will find it difficult to convey, and the student may find it difficult to understand, these topics unless their mutual visualization of three-dimensional problems is good or if their grasp of the mathematics and nomenclatures of symmetry and symmetry operations is far better than it is now. The use of models and more elaborate visual aids will become essential.

It has always been hoped that one of the consequences of a developed understanding of the mechanisms of inorganic reactions would be the availability of the so-called planned synthetic path (characteristic of organic synthesis). Apart from certain areas which have been discussed in previous chapters this ambition is as far away as ever and much that goes under the heading of planned synthesis is really interpolation or extrapolation based on past experience. Far too many discoveries of new types of compound are still serendipidous in character. The reaction mechanisms of boron and the boranes have been deliberately omitted from this book, partly because the stage has not yet been reached where a simple survey at this level will be of any value. However, the extent of development in the structural and preparative chemistry in this area indicates quite clearly that the study of the hydrides of carbon and their derivatives (that is, organic chemistry) will never compare in its richness of variation of coordination number, geometry, and bond type with the chemistry of the hydrides of boron and their derivatives. The mechanistic studies, when they reach the level of understanding at present attained for carbon alone, will be rich and challenging indeed.

When it comes to making predictions about the wider application of this sort of knowledge anyone who has any confidence in his predictive abilities should spend his time writing patent applications rather than books. However, it is clear even to a distant observer that the move is away from the empirically brewed catalyst mixtures towards fairly well-understood and fairly simple single-compound catalysts, many of which fall into the category of complex inorganic or metal-organic species. People speak of the 'tailoring of catalysts' (by variation of a ligand here and a ligand there) to promote a very specific reaction. In this way we move very slowly in the direction of enzyme catalysis, but we are still a long way off. It is likely that these organo-metallic catalysts will provide one of the pathways to achieving the chemical engineer's dream of moving away from the sledgehammer techniques of high temperatures and pressures with poorly understood heterogeneous catalysts, towards mild reaction conditions where the catalysts are active and easily reproduced. These catalysts will be very specific and require fewer processing steps. They will function effectively on cheap, readily available, and reasonably inactive feed stocks such as natural hydrocarbons, water, oxygen, and nitrogen. The reduction in the need for the expensive plant for high pressures and temperatures will certainly reduce capital cost, the greater cleanness of the reactions may very well reduce the amount of thermal and chemical pollution that should be a source of shame to the industry. Processes to convert waste products back into useful intermediates will replace dumping of rubbish if, and only if, they become cheaper than dumping; the use of homogeneous catalysts in this area ought not to be discounted.

It used to be said that the prosperity of a country could be reflected in its figures for the production of sulphuric acid, but that was in the days of coal and steam and battleships and T.N.T. It is likely that in the future the better figure will be the figures for the utilization of homogeneous organo-metallic catalysts.

Although biological aspects of inorganic complexes have been completely left out of this book it should be pointed out that the future holds promise not just for an increase in the chances given to the scientists to 'play God' and design and study simple organo-metallic chemical models as a means of understanding the complex and highly specific biological systems, but also for a new pharmacopeia. Certain optically active octahedral tris-chelate transition metal cations can be very effective anti-virus or anti-bacterial agents, even though the coordination shell of the metal is unchanged *in vivo*; certain simple platinum(II) complexes can be used as effective anti-tumour agents and the use of potential polydentate ligands as a means of regulating the concentration of metal ions (for example Cu^{2+}) in certain metabolic or excretory disorders has been successful. One should expect extensive developments in these areas.

The book has set out to deal with inorganic reaction mechanisms and yet this last chapter has dealt mainly with reactions at carbon. Perhaps this is the most obvious sign for the future because we must surely then escape from this hang-over of the 'Vital Force' theory which still divides the subject into the separate artificial compartments labelled organic and inorganic chemistry.

Problems

11-1 Distinguish between the terms, *isotactic, syndiotactic,* and *atactic*. What is the principle difference between a catalytic path that leads to the formation of isotactic polypropylene and one that leads to the formation of syndiotactic polypropylene.

11-2 The Wacker process involves the oxidation of ethylene to acetaldehyde in the presence of aqueous Pd(II) chloride. Given that the reduction of palladium follows the rate law,

$$\text{rate} = \frac{k[\text{Pd(II)}][\text{C}_2\text{H}_4]}{[\text{H}^+][\text{Cl}^-]^2}$$

write a plausible mechanism for the process

$$\text{PdCl}_4^{2-} + \text{C}_2\text{H}_4 + \text{H}_2\text{O} \rightleftharpoons \text{CH}_3\text{CHO} + \text{Pd} + 4\text{Cl}^- + \text{H}^+$$

Note that, if the reaction is carried out in D_2O, no deuterium is incorporated in the product.

Bibliography

Bawn, C. E. H. and A. Ledwith. Stereoregular addition polymerization. *Quart. Revs.*, 1962, **16**, 361.

Candlin, J. P., K. A. Taylor, and D. T. Thompson. *Reactions of Transition Metal Complexes*. Elsevier Publishing Company, Amsterdam, 1968.

Heck, R. F., Insertion reactions of metal complexes. *Advances in Chemistry Series*, No. 49, p. 181, A.C.S., 1965.

Henrici-Olivé, G. and S. Olivé. Non-enzymatic activation of molecular nitrogen. *Angew. Chem.*, Int. Ed., 1969, **8**, 650.

Henrici-Olivé, G. and S. Olivé. Influence of ligands on the activity and specificity of soluble transition metal catalysts. *Angew. Chem.*, Int. Ed., 1971, **10**, 105.

Jones, M. M., *Ligand Reactivity and Catalysis*. Academic Press, New York, 1968.

Index

Activation, free energy of, 8
Addition reactions, 12, 25
Addition-rearrangement, mechanism of oxidative addition, 155
Adiabatic electron transfer, 131
Allogons, 115
Amines, nucleophilicity of, 52
Anation, 90
Aquation, steric course, 103
Arsines, nucleophilicity of, 52
Atactic polymers, 174

Bailar twist, 121
Base hydrolysis, 94–96
   steric course, 103
Berry twist, 111, 115–119
Biphilic reagents, 53
Bonding in d^8 complexes, 45
Bridge transfer, 134
Bridgehead reaction centres, substitution reactions at, 35

Carbon
   nucleophilic substitution at, 28–32
   stereochemistry of substitution at, 29–32
Carbonium ions, 28, 30, 93
Catalysis, 168–187
   of redox reactions, 148
Chromium(II)
   oxidative-addition reactions, 161
   reduction by, 133, 136, 138, 147
Chromium(III)
   catalysed substitution reactions, 164
   reduction of, 137, 139–142
   substitution reactions, 80, 85, 87, 88, 91
Cis effects, 58, 88
Cobalt(I)
   catalysts in hydroformylation, 182
   as a nucleophile, 157
   oxidative-addition reactions, 154, 156
Cobalt(II)
   reduction by, 129, 136–138, 142
   substitution reactions, 38, 39
   oxidative-addition of cyano complexes, 161

Cobalt(III)
   reduction of, 129, 133–142
   stereochemistry of substitution, 103–106, 108–110
   substitution reactions, 84–106
Complementary reactions, 145
Complex formation, kinetics of, 82
Conjugate-base mechanism, 95
Coordinated ligands, reactions of, 13
Coordination number, dependence on electron configuration, 152
Coordination unsaturation, 44, 152, 155
Cryptosolvolysis, 97
Crystal-field theory, application to reactivity, 77, 82

d^8 configuration, coordination number and geometry, 44
Dichromates, substitution reactions, 38
Dinitrogen complexes, 163, 170
Dioxygen complexes, 156
Dissociative interchange, 83
   distinction from dissociative mechanism, 92, 100
Dissociative mechanism, distinction from I_d, 92, 100

Electron transfer, 125
Electron transfer to bridge, 139
Electron tunnelling, 131
Elimination reactions, 12, 25

Five-coordinate complexes
   occurrence, 69, 79
   substitution reactions, 69–75
Franck–Condon restrictions, 129

Germanium compounds, reaction mechanisms, 32
Gold(III)
   comparison with Pt(II), 60
   catalysed substitution reactions, 163
   nucleophilicity scale, 54
   substitution reactions, 45, 54, 60, 62

Halides, nucleophilicity of, 52

Hydrogenation of alkenes, 177–182
Hydroformylation, 15, 182–186

Inertness, 6
Inner-sphere redox reactions, 133–142
  differentiation from outer sphere, 133, 142
  rate determining bridge formation, 134
  slow bridge fission, 138
Insertion reactions, 14, 172
Intermediates
  distinction from transition states, 19
  five-coordinate in octahedral substitution, 84, 96, 104
  in substitution reactions, 19
Intramolecular isomerization, 109, 110–122
Intrinsic reactivity, 53
Inversion
  in 3-coordinate systems, 112
  in 4-coordinate systems, 114
  at tetrahedral carbon, 29
  at tetrahedral silicon, 34
Ion-association, 91, 97, 99, 101
Ion-pair formation, 91, 97, 100
Iridium(I), oxidative-addition reactions, 152, 154, 156, 158, 160
Iridium(III), substitution reactions, 87
Iron(II)
  reduction by, 128, 130–132, 136, 140
  substitution reactions, 39
Iron(III)
  reduction of, 128, 130–132, 135, 138, 145, 148
  substitution reactions, 81, 82
Isomerization
  of octahedral complexes, 108–111
  of square planar complexes, 64, 66, 114
Isotactic polymers, 174

Kinetics, 6

Labilizing effects
  in octahedral substitution, 88
  in square-planar substitution, 54
Ligand migration, 14, 172
  stereochemical evidence, 173
Linear free energy relationships, 51, 88
Linkage isomerization, 102

Mechanism, determination of, 4–10
Molecularity
  Ingold definition, 18
  Langford–Gray nomenclature, 20

n_{Pt}, 51
Nickel(II)
  stereochemistry and coordination number, 44
  substitution reactions, 39, 54
Nickel carbonyl, substitution reactions, 39
Nitrogen fixation, 169–171
Nitrosyl complexes
  oxidative addition, 153
  substitution in tetrahedral, 40
Non-complementary reactions, 145
Nucleophilic discrimination factor, 52, 59
Nucleophilic substitution, mechanism of oxidative addition, 156
Nucleophilicity scales, 51, 54

Octahedral substitution, 76–107
  associative mechanism, 84
  the D mechanism, 84, 92, 96, 102
  the I_d mechanism, 82, 91, 97, 100
  the nature of the leaving group, 87
  in non-aqueous solvents, 96
  role of the central atom, 87
  steric course, 103–106
  typical mechanism, 83
Olefin complexes, 155, 172–184
Oligomerization, 175–177
Outer-sphere complex formation, 82
Outer-sphere redox mechanism, 127–133
  differentiation from inner-sphere, 128, 142
Oxalate, oxidation of, 11, 146
Oxidation and reduction, 124–150
Oxidation state, 12, 124
Oxidative addition, 152–162
  kinetics, 154
  mechanisms, 153
  one-electron processes, 161
  stereochemistry of, 158–161
Oxy-anions, reactions of tetrahedral transition metal, 38

Palladium(O) complexes, 39, 153, 154
Palladium(II), substitution reactions, 54, 62–64, 66
Phosphines, nucleophilicity of, 52
Phospholanes, 37
Phosphorus(V), pseudorotation, 115–119
Phosphorus compounds
  reaction mechanisms, 36
  stereochemistry of substitution, 36
Platinum(O) complexes, 39, 153, 154, 155

Index

Platinum(II)
  comparison with Au(III), 60
  five-coordinate complexes, 71, 73
  nucleophilicity scale, 51
  oxidative-addition reactions, 158
  substitution reactions, 44, 47–67
Platinum(IV), catalysed substitution reactions, 165
Polymerization of olefins, 171–177
Proton basicity, relationship to nucleophilicity, 52
Pseudorotation, 111
  in 4-coordinate systems, 114–115
  in 5-coordinate complexes, 115–119
  in 5-coordinate Pt(II) complexes, 67
  in 6-coordinate complexes, 119–122
  in high coordination number complexes, 121
  of hydrido complexes, 122

Racemization, 32, 110
Ray and Dutt twist, 121
Reaction type, classification of, 11, 22
Redox catalysed substitution, 163–166
Redox reactions, 12, 124–150
  number of electrons transferred, 143
Reductive-elimination, 162
Reppe catalysts, 175
Resolution of phosphines, 113
Rhodium(I), hydroformylation catalysts, 183
  hydrogenation catalysts, 172–182
  oxidative-addition reactions, 153
  substitution reactions, 62
Rhodium(III), substitution reactions, 80, 85, 87, 91, 92, 94
Ruthenium(III)
  reduction of, 135, 138, 140–142
  substitution reactions, 87

Signposting, 32
Silicon compounds
  substitution reactions, 32
  stereochemistry of substitution, 34
Solvated electrons
  reactions of, 125
  reduction of metal complexes, 126
Solvated metal cations, substitution reactions, 80–83
Solvation, 5
Solvento intermediates, 47
Solvolytic reactions, 86
Square-planar complexes
  occurrence, 42
  isomerization reactions, 64, 66, 114

Square-planar substitution
  bond making–bond breaking relationships, 61
  characteristic mechanism, 46–48
  dissociative mechanism, 63, 67
  factors influencing reactivity, 50
  first-order path, 47
  geometry of transition state, 48
  kinetics, 45
  leaving group effects, 59
  nature of the reaction centre, 60
  steric course, 50, 66
  substitution reactions, 42–68
Stability, 6
Stereochemical change, 108–123
  without substitution, 109
Stereochemistry
  substitution at carbon, 29–32
  octahedral substitution, 103–106
  substitution at phosphorus, 36
  substitution at silicon, 34
  substitution at square-planar complexes, 50, 66
Steric hindrance, in Pd(II) and Pt(II) complexes, 47, 54, 63
Substitution
  associative, 19, 21
  bimolecular homolytic, 17
  definition, 16
  dissociative, 19, 21
  electrophilic, 17
  general considerations, 16–25
  interchange, 21
  nucleophilic, 17
  redox catalysed, 163–166
  unimolecular homolytic, 16
Syndiotactic polymers, 174

Template reactions, 14, 177
Tetrahedral carbon, substitution reactions, 27–32
Tetrahedral compounds
  occurrence, 25
  substitution reactions, 25–41
Tetrahedral substitution
  at d^{10} complexes, 39
  transition metal complexes, 38, 39
Thallium(II), identification as intermediate, 148
Thioethers, nucleophilicity of, 52
Tin compounds, substitution reactions, 32
Tin(III), identification as intermediate, 146
Trans effect, 54–58
  definition, 55

ground state, 56
in octahedral complexes, 88
quantitative measurement, 55
sequence, 55
transition state, 57
Transition state, 7
for square planar substitution, 48
Trigonal prismatic complexes, 120

Valence-bond theory, application to reactivity, 77

Vanadium(II)
reduction by, 135
substitution reactions, 81
Vanadium(III), substitution reactions, 85
Vitamin B_{12}, 157

Walden inversion, 29
Water exchange, rate constants, 81

Ziegler–Natta catalysts, 172–175